Planning for Survivable Networks:
Ensuring Business Continuity

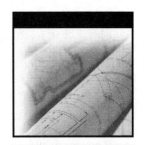

Planning for Survivable Networks:
Ensuring Business Continuity

Annlee Hines

Wiley Publishing, Inc.

Publisher: Robert Ipsen
Editor: Carol A. Long
Developmental Editor: Adaobi Obi
Managing Editor: Micheline Frederick
Text Design & Composition: Wiley Composition Services

Designations used by companies to distinguish their products are often claimed as trademarks. In all instances where Wiley Publishing, Inc., is aware of a claim, the product names appear in initial capital or ALL CAPITAL LETTERS. Readers, however, should contact the appropriate companies for more complete information regarding trademarks and registration.

This book is printed on acid-free paper. ∞

Copyright © 2002 by Annlee Hines. All rights reserved.

Published by Wiley Publishing, Inc., Indianapolis, Indiana
Published simultaneously in Canada

No part of this publication may be reproduced, stored in a retrieval system, or transmitted in any form or by any means, electronic, mechanical, photocopying, recording, scanning, or otherwise, except as permitted under Section 107 or 108 of the 1976 United States Copyright Act, without either the prior written permission of the Publisher, or authorization through payment of the appropriate per-copy fee to the Copyright Clearance Center, Inc., 222 Rosewood Drive, Danvers, MA 01923, (978) 750-8400, fax (978) 750-4470. Requests to the Publisher for permission should be addressed to the Legal Department, Wiley Publishing, Inc., 10475 Crosspointe Blvd., Indianapolis, IN 46256, (317) 572-3447, fax (317) 572-4447, E-mail: permcoordinator@wiley.com.

Limit of Liability/Disclaimer of Warranty: While the publisher and author have used their best efforts in preparing this book, they make no representations or warranties with respect to the accuracy or completeness of the contents of this book and specifically disclaim any implied warranties of merchantability or fitness for a particular purpose. No warranty may be created or extended by sales representatives or written sales materials. The advice and strategies contained herein may not be suitable for your situation. You should consult with a professional where appropriate. Neither the publisher nor author shall be liable for any loss of profit or any other commercial damages, including but not limited to special, incidental, consequential, or other damages.

For general information on our other products and services please contact our Customer Care Department within the United States at (800) 762-2974, outside the United States at (317) 572-3993 or fax (317) 572-4002.

Wiley also publishes its books in a variety of electronic formats. Some content that appears in print may not be available in electronic books.

Library of Congress Cataloging-in-Publication Data:

ISBN: 0-471-23284-X

Printed in the United States of America

10 9 8 7 6 5 4 3 2 1

For Eric and Aylyffe

sine qua non

Contents

Foreword		**xiii**
Chapter 1	**Introduction**	**1**
	Network Continuity	2
	Define *Survival*	3
	In Defense of Paranoia	4
	By the Numbers	5
	Borrow from Einstein	6
	Think the Unthinkable	8
	Plan to Survive	8
	Choice versus Chance	10
Chapter 2	**Network Threats**	**11**
	Kinds of Attacks	12
	Immature Hands	13
	Voyeurs	14
	Testers	18
	Deliberate Attackers	19
	Mature Hands	26
	Industrial Espionage	27
	Fraud/Theft	29
	Record Alteration	31
	Extortion	33
	Externalities	33
Chapter 3	**Tactics of Mistake**	**35**
	TCP/IP	36
	Probes	43
	Viruses	45

	Worms	46
	Trojan Horses	48
	Denial of Service/Distributed DoS	49
	Sample Attack	51
	Means	55
	Opportunity	56
Chapter 4	**Murphy's Revenge**	**57**
	System Is Not a Dirty Word	57
	Complexity	58
	Interaction	58
	Emergent Properties	59
	Bugs	59
	Where Opportunity Knocks	60
	Top General Vulnerabilities	61
	#1: Default Installations	61
	#2: Accounts with Weak/No Passwords	62
	#3: Nonexistent or Incomplete Backups	63
	#4: Large Numbers of Open Ports	63
	#5: Not Filtering for Correct Ingress/Egress Addresses	64
	#6: Nonexistent or Incomplete Logging	64
	#7: Vulnerable CGI Programs	65
	Top Windows Vulnerabilities	65
	#1: Unicode Vulnerability	66
	#2: ISAPI Extension Buffer Overflows	66
	#3: IIS RDS Exploit	66
	#4: NETBIOS—Unprotected Windows Networking Shares	66
	#5: Information Leakage via Null Session Connections	67
	#6: LM Hash	67
	Top UNIX Vulnerabilities	68
	#1: Buffer Overflows in RPC Services	68
	#2: Sendmail Vulnerabilities	68
	#3: BIND Weaknesses	68
	#4: r Commands	69
	#5: LPD	69
	#6: sadmind and mountd	69
	#7: Default SNMP Strings	70
	Common Threads	70
	Design Your Way Out of Trouble	72
	Topology	72
	Physical Topologies	72
	Logical Topologies	73
	Defense in Depth	75
	The Price of Defense	78
	Olive-Drab Networks	80
	Benefits	80
	Costs	81

	Converged Networks	82
	The Catch	84
	Operator Error	85
Chapter 5	**"CQD ... MGY"**	**87**
	A Classic Disaster	88
	Lessons from Failure	90
	A Trophy Property	90
	Warning Noted …	92
	Train the Way You Will Fight	92
	What Did You Say?	93
	A Scarcity of Heroes	94
	Lessons from Success	94
	Organization	95
	Training	96
	Attitude	97
	A Plan	98
	What Are You Planning For?	99
	Adequate Warning	99
	Not Just Hurricanes	102
	Major Storm Effects	103
	Modest Warning	105
	No Real Warning at All	107
	It's a Scary World, Isn't It?	113
Chapter 6	**The Best-Laid Plans**	**115**
	Three Main Points	115
	Operational Continuity	116
	Twenty Questions	117
	A Few More Questions	122
	Getting the People Out	124
	Off-Site	124
	On-Site	125
	Network Assets	126
	Example: Data Services	129
	Lessons Actually Learned	135
	Topology	136
	Facilities	136
	Configuration Control	136
	The Right Tools for the Job	137
	Lessons Potentially Learned	138
	Kudos	138
	Extending the Example	139
Chapter 7	**Unnatural Disasters (Intentional)**	**143**
	Physical Attacks	146
	Bombs	147
	Electromagnetic Pulse	147

        Sabotage 148
        CBR Attacks 149
    World Trade Center Examples 153
        Successes 154
            NYBOT 154
            The Wall Street Journal 156
            Lehman Brothers 158
        Lost Access 159
        Less Than Successes 162
            The Local Loop 162
            New York City OEM 164
            The U.S. Secret Service 165
    Cyber-Attacks 166
        Cyber-Kidnapping 166
        Extortion 167
        Easier Targets 167
    Combined Attacks 168

Chapter 8 Unnatural Disasters (Unintentional) 171
    Unfortunate Opportunities 171
        Reportable Outages: They're Everywhere 172
        Route Diversity in Reality 175
        Fire 175
        Required Evacuations 178
    Unfortunate Planning 178
        Yours 178
        Theirs 181
    Unfortunate Implementation 186
        Equipment 1, Plan 0 186
        Solving the Wrong Problem 188
            Candidates 188

Chapter 9 Preparing for Disaster 191
    Define *Survival* 191
        What Must Roll Downhill 192
    Survival Requirements 194
        Network Continuity Requirements 195
        Threat Analysis 202
            Physical Threats 202
            Cyber-Threats 204
        Operational Analysis 206
    Survival Planning 207
        Fixes 207
        Remedies 210
        Procedures 211

	Survivability Today	213
	Don't Get Too Close	214
	Talk Is Cheap	215
	Data Currency	217
	Trade-offs	218
Chapter 10	**Returning from the Wilderness**	**219**
	Cyber-Recovery	220
	Operational Procedures	220
	Forensic Procedures	221
	Physical Recovery	226
	Immediate Operations	226
	Sustained Operations	227
	Restoration	228
	Undress Rehearsal	231
	Exercise Scenario 1: Cyber-Problems	234
	Exercise Scenario 2: Physical Problems	235
	Evolution	236
Chapter 11	**The Business Case**	**243**
	Understanding Costs	244
	Fixed and Variable Costs	244
	Direct Costs versus Indirect Costs	245
	Explicit and Implicit Costs	247
	Valid Comparisons	247
	Understanding Revenues	249
	Expected Values	250
	Presenting Your Case	252
	CDG Example	255
	Alternatives Considered	256
	Disaster Summary	256
	Alternatives Summary	259
	Risks Not Mitigated	260
	Finally	262
Chapter 12	**Conclusion**	**263**
	Necessity	264
	Basic Defenses You Must Implement	265
	The Deck Is Stacked Against You	266
	Catastrophes Happen	267
	Your Recovery	268
	Trade-offs	270
	Systemic Behavior	270
	Standardization versus Resiliency	272
	Pay Me Now or Pay Me Later	273

Appendix A	**References**	**275**
Appendix B	**Questions to Ask Yourself**	**281**
Appendix C	**Continuity Planning Steps**	**285**
Appendix D	**Post-Mortem Questions**	**289**
Appendix E	**Time Value of Money**	**291**
Appendix F	**Glossary**	**293**
Index		**299**

Foreword

It is a mistake to try to look too far ahead. The chain of destiny can only be grasped one link at a time.
Winston Churchill

It's true that the events of September 11, 2001 crystallized my thoughts about network survivability, but the thoughts go back much further than that. I became very interested in terrorism while serving in the USAF in Europe, where it was a very real threat, especially to those of us in an American uniform. That interest had been somewhat dormant, but it never really went away. I stayed aware of the threats and how they were evolving; where once terrorists struck only where they could melt away into the populace to live and strike another day, they no longer care about that. This is a watershed, for it changes the nature of the threat: Delivery need no longer be safe for the deliverer. That turns previously untouchable locations into targets.

Since I left the service, I have become a network engineer after owning two businesses, and the bottom-line responsibility I held there changed the way I thought about business; it has also affected how I look at network operations. The network exists only because it brings value to its business. But if it brings value, that value must continue or the business itself may suffer such a degradation of its financial condition that it is in danger of failing. That statement was not always true, but it has become true in the past two decades. Almost unnoticed, networks have indeed become integral to the operations of all major businesses, all around the world.

What is more, we do operate in a global economy, with costs held to their barest minimum in the face of competition from other companies, some of whom operate in other countries, where cost structures are different. If the network is a major factor in your firm's competitiveness, whether from a perspective of increasing productivity or a perspective of minimizing the cost of timely information transfer, its continuity is critical to business continuity.

The networking community was as mutually supportive as ever during and after the terrorist attacks of September 11. The NANOG (North American Network Operators Group) mailing list was flooded with advisories of where the outages were, who was able to get around them, offers of available bandwidth and even temporary colocation, if needed. There were also dire thoughts concerning how much worse the situation would have been had a couple of other locations been hit as well.

Many of the first responders who died lost their lives due to communications failures—they did not notify the command center of their presence or location, but rushed in to help because lives were at stake right now. When the command center decided to evacuate because senior officials knew the buildings could not stand much longer, radio coverage was so spotty that some who lost their lives did so because they simply never got the word to get out. The communication network that day was inadequate to the task.

After the collapse of the World Trade Center, much of the information dissemination was made via email and Internet; those hubs were the ones referred to on NANOG in the what-if discussions. Networks have always been about communications—moving the information from where it is already known to where it needs to be known to add value. "Rejoice! For victory is ours," gasped Phaedippides with his dying breath after running from the battlefield at Marathon to Athens. His message had value because Athens expected to lose the battle, and the city fathers were preparing to surrender when they saw the Persian fleet approach.

On a more business-centric note, the time to buy, said Lord Rothschild, is when the blood is running in the streets. He used his superior communications to cause that to happen, after the Battle of Waterloo, and he made a financial killing in the London markets his better information had manipulated.

Your network is the nervous system of your business—the connector between its brains and direction and the actual execution of business decisions. If the nervous system is damaged or disrupted, bad decisions may ensue (from bad information), or good decisions may be ordered but never executed. Either way, it might be your company's blood that is running in the streets.

Business continuity implies that the organization continues to operate as a business; for this, the nervous system must continue to be there. It may not be there in all its ordinary glory, but the essential services it provides must continue to be present. Getting those defined and finding ways to ensure their continuity are the subjects of this book.

The threats to continued network operation range from the dramatic (major terrorist attack) through the more common, but still not frequent (natural disasters), to the threat attacking you every day (hackers). The tools that protect you from the first two are quite similar; there is also considerable overlap with the tools to protect you from the third. As with anything in either networking or business in general, you are going to have to make compromises. If you learn from the principles addressed here, rather than blindly answering the lists of questions presented, you will be prepared to make the hard choices on a knowledgeable basis. They won't be any more pleasant, but the consequences are less likely to be an unpleasant surprise.

No book ever springs full-blown from the author's forehead, like the fully armed Athena. I have had so much help I cannot begin to thank those people. From years ago, I owe Colonel Richard W. Morain (USAF, Retired) for his patience and support. Even after I left the service, he maintained contact, and I am better for it. More recently, I've wound up doing this through the intervention, after a fashion, of Howard W. Berkowitz, who liked my comments on a mailing list, and offered me the opportunity to write about networking for publication. Then it was a review of his manuscript that put me in contact with his editor, Carol Long, at John Wiley and Sons.

During the hashing-out process of what this book would actually become, and the grind of getting it all down in bytes in a lot of files, Carol's support has been invaluable. Likewise, my friends at Nortel have maintained an enthusiasm for the project when my energy flagged; chief among them have been Ann Rouleau and John Gibson. My manager at the time, Mark Wilson, massaged the administrative system to propitiate the intellectual-property gods; he had more patience than I, lots of times.

And, of course, *sine qua non*, have been my family, who now expect me to do this again. With their help, I will.

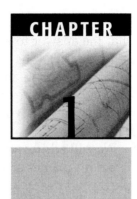

CHAPTER 1

Introduction

It's choice, not chance, that determines your destiny.
John Nidetch

I felt the explosion through the building as much as I heard it. The next sound was glass crashing to the sidewalk below, clearly audible because the windows of the office in which I stood had blown out along with all the rest. I remember hoping no one was on the sidewalk to be hit by all that—I even stepped over to look, then I was out of the office, down the stairs, and into my own office, securing the classified documents with which I had been working and helping the two others in early that day to secure all their classified material.

This wasn't September 11, 2001; it wasn't the World Trade Center or the Pentagon. Rather, this bombing occurred at HQ USAFE/AAFCE, at Ramstein AB, Germany, 20 years before. It wasn't done by al Qaeda, either; they didn't invent terrorism. As noted thousands of years ago in Ecclesiastes, there is nothing new under the sun. The latest terror attacks against Americans are bigger, and they are on our soil, but they are not a truly new phenomenon.

That does not mean they are not new to you or to other individuals. Nonetheless, as a society, we already know what we have to do. It's now a matter of making the effort and spending a little money.

How much you need to spend depends on your circumstances (but isn't that the answer to every question about networks?). What you need will not necessarily be what your closest competitor needs, not to mention what someone in an entirely different business needs. The first thing you need, whatever your business may be, is good information and an understanding of what you need to protect as well as what you are protecting it against.

We will address those questions in the course of this book. Nonetheless, it is not a primer or a checklist for how to do this task or that. For the most part, you already know how to do your job, whether you are the CIO or a senior network administrator/architect responsible for a global corporate network. The networking world has changed since September 11, 2001, and you have to reconsider how vulnerable the information nervous system of your company is.

Network Continuity

Business continuity is a subject that has been around for a while, and governmental continuity is not new, either. Businesses routinely restrict the travel of critical personnel; no more than so many of the senior leadership (sometimes no more than one) may travel on the same flight, for instance. One disaster, or even one mischance, cannot leave the company without leadership.

Likewise, democratic governments have standard lines of succession publicly preestablished, and the entire designated line of succession is never together in one convenient location, to be removed by one mad act or catastrophe. When the President of the United States visits the Capitol to address a joint session of the Congress, the entire line of succession could be present: The Vice President is also the President of the Senate and normally co-presides with the Speaker of the House of Representatives, who is next in line. Next in the designated succession is the President pro tempore of the Senate, who (as a senior member of the legislative branch) is also present, followed by a succession sequence established by law from the President's cabinet.

One cabinet secretary, at least, is always missing, designated to not attend and thereby be available to preserve the continuity of the U.S. government. It is a dubious honor at best; the media often assumes it is someone who drew the short political straw. Perhaps, on some occasions, it is. But when terrorists did strike inside the United States, the Presidential succession was immediately dispersed and remained dispersed until security functions believed the likelihood of a decapitating blow was no longer present.

Your company's senior leadership probably does not have such drastic measures preplanned and implementable at a moment's notice. Nor, frankly, is a civilian business likely to need to secure the persons of key decision makers. No matter how large your business, it is not in a position to do what the leader of a Great Power can do with just a few words.

Your company's senior leadership does make real decisions, with substantial consequences, every day. Those decisions are only as good as the quality of the information on which they are based. Good information—information that has accuracy and integrity and that is available where and when it is needed—does not come from the Tooth Fairy, nor does it come from the good intentions of honest people working very hard. It may be created by such people, but it will be delivered to those who need it, when and where they need it, only by a network that is available, reliable, and trustworthy: a survivable network.

Making that happen, despite natural and unnatural disasters, despite the inevitable mistakes of well-intentioned, honest people, and despite the disruptions of skilled and semi-skilled cyber-vandals, is network continuity.

Define *Survival*

On a fundamental, physical level, *survival* is a simple thing: staying alive. That does not necessarily mean staying fully functional, or even partially so, unless you modify the definition to include some performance characteristics.

What does survival mean to your company?

You cannot define what survival means to your network until you know what it means to your company. The network serves a business purpose; without that purpose it would not exist. What is your company's core function, the function without which it would cease to exist? Must it continue doing that very thing, or does it, in fact, do something more fundamental that could be done in a different way from how you do it now?

If that seems a little confusing, step back and look at your company from the vantage point of your customers. You manufacture and sell books, perhaps, like John Wiley and Sons, the publisher of this book. What are your customers buying when they buy your books? Black, or even colored, scratches on processed wood pulp have no value. *Content* has value; customers are buying the information contained in the book. This ink-and-paper delivery vehicle is convenient enough, and we are all certainly used to it and know how to deal with it, but it is hardly the only means of delivering information to a paying customer.

That's a lovely sound to someone with profit-and-loss responsibility: *paying customer*. The key to knowing what survival means to your business is to know what your customers are paying for. That is, not what you or your CEO or your Board of Directors thinks the customers are buying, but what the customers think they are buying. If your business can continue to provide that, whatever *that* is, despite the slings and arrows of outrageous fortune, then your business will survive. It is up to the senior management and Directors to understand what *that* is. They will pass their understanding to you in the form of the business operating characteristics that must continue.

What will it take for your network to support them?

In Defense of Paranoia

What are you afraid of concerning your network? What *should* you be afraid of? Those are not necessarily congruent sets. September 11, 2001, made us all aware of terrorism and of the threat of airplanes being used as bombs to destroy buildings.

How often has that actually happened? Once. Horrific as it was, involving four separate aircraft, as an event it has happened only once. Some businesses located in the World Trade Center will not survive; they simply lost too much. Others continued to operate with hardly a noticeable ripple to their customers. Most muddled through somewhere in between. It is not fair to say that our military headquarters was unaffected, for it surely was. Military information systems, though, were robust enough to avoid serious disruption to any of the command and control functions—the networks delivered, with a little help from the human elements. We will examine a few exemplary stories from the attack on the WTC (civilian networks are more directly comparable for our purposes); in these cases, the companies' networks were prepared, some better than others, and they continued to deliver the business for their companies. There are other examples, not as positive, that we will examine, as well. We do well to remember Santayana: "Those who cannot remember the past are condemned to repeat it."

Far more common than terror attacks are natural disasters. Hurricane Andrew, a Category 5 storm, devastated the southern end of Florida, and some areas have simply never recovered. A few years later, Hurricane Hugo, a Category 4 storm, swept through the Carolinas and wreaked substantial destruction there. California has suffered two major earthquakes in the past 13 years: Loma Prieta, in 1989, and Northridge, in 1994. As in all other major natural disasters, basic utilities were disrupted, in some areas for a surprisingly long time. The Kobe-Osaka earthquake in Japan in 1995 was even stronger (damage estimates reached 2 percent of the area's Gross Domestic Product). Devastating tornadoes strike cities in the United States every year. Mount St. Helens' eruption in 1980 devastated a large area of Washington, not with a lava flow, but with pyroclastic flows and lahars; they were far from the first such flows and lahars in the Pacific Northwest's history. The same is true of Mount Pinatubo in the Republic of the Philippines; the eruption in 1991 caused massive destruction in the surrounding area.

Should you be more concerned about natural disasters than unnatural ones—those caused by your fellow man? Yes and no. Some unnatural disasters are not deliberate; they occur because humans are sometimes sloppy or lazy in their work, and sometimes they are ignorant of the consequences of a particular action. Urban floods are not always an act of nature; sometimes they are the intersection of digging equipment and a major water main (or even, as in Chicago, the underground side of a river).

Fortunately, your preparations to deal with natural disasters form a good foundation for preparation to deal with a terrorist attack. In both cases, you are preparing to lose the use of a major networking location for an indeterminate period of time. You are concerned about saving your people first—equipment is far easier to replace, and arrangements can be made quickly for new desktops and servers, new routers and switches. Arrangements for a new operating location may prove more difficult; that will depend on the magnitude of the disaster and the condition of the local real estate market at the time. Your planning can mitigate even that.

Natural disasters are your first priority; with a security twist, that planning will ensure network continuity, right?

Wrong.

Wrong, wrong, wrong.

By the Numbers

By far, the most common attack your network will endure is one that it has already endured, probably more than once. I am not being rude when I say that I hope you noticed.

Cyber-attacks come early, and they come often. They are also characterized by enormous variability, which makes them harder to defend against. Some are a sledgehammer, taking down entire segments and rendering them inoperable for far too long. Some are mere probes, testing to see if you noticed. And some are subtle, slipping in, extracting information (perhaps altering it as well), and slipping away without doing anything to attract notice.

You must also defend against cyber-attacks with one hand tied behind your back: The protocols on which your network depends are grievously insecure. They were designed in a time when only a few academic and some trusted government agencies needed to interconnect computers. Everyone knew everyone else, and the goal was to create openings from one system into another in order to share information.

Networking has evolved quite a bit from that.

Now your task is far more likely to be to prevent access by unauthorized users than it is to make information available to anyone who asks. Everyone and his hacker nephew, it seems, has a computer and access to the Internet. Your business needs access to *and from* the Internet in order to conduct business and to obtain and move the information needed to create value for your customers. You must facilitate the readiest possible access from the inside out, so that your company's employees can do their jobs, and establish carefully-controlled-yet-easy-for-the-customer access from the outside in. And you must do this in the most economical way possible because you, quite possibly, do not directly contribute to your company's revenues. (That's a polite way of saying you're a cost center.)

If anyone can get in and muck about with your data, how can those senior managers who must make decisions have confidence in the choices they make? If necessary, how could they defend their choices in court if those choices were flawed at the foundation?

You probably know all this already, though you haven't articulated it beyond muttering into your coffee on occasion. But now *is* the occasion, and you should do more than just mutter into your coffee. Thanks (if that is an appropriate word) to the events of September 11, 2001, senior management teams and Boards of Directors have realized that business continuity is about more than travel restrictions and who will succeed the CEO if he has a heart attack.

Borrow from Einstein

The current climate of reassessment is one in which a carefully presented plan to ensure network continuity, in support of business continuity, can gain approval and that all-important follow-on to approval: funding. As you will see in one instance at the World Trade Center, an approved continuity plan that is not funded may as well have never existed.

The same is true of implementation. Once you have an approved plan and funding earmarked, you must not let the funding be diverted for anything else. You must be especially sensitive to raids, as it were, on your operating budget because you have this "extra" money at your disposal.

To help you protect your budget for preserving network functionality, you may need to borrow from the techniques of Albert Einstein, among other great scientists. To explain difficult theoretical concepts, he sometimes used what are called *thought experiments*. This is a very clever term, for science, as we all know, is very fond of experiments to validate a given theory. Thought experiments ask the participant to imagine what happens in a certain situation, based on everyday experience. Because we largely understand how the universe works, we can imagine an outcome that we are confident maps to reality.

Here is your thought experiment to protect (or obtain and then protect) your funding for network continuity:

How would *<insert company name here>* do business without the network?

At this point in your company's life, the better question might well be "*Can* the company do business without the network?" Theoretically, the answer is yes because business per se is as old as history. But consider your profit margins (quite possibly thin) and your cost structures. Reduce productivity by how much people use the network to obtain and exchange information. If you have no real measurements for this (and few people do), use a naive figure of 50 percent. Could your business still earn a profit in today's market with 50 percent of your current productivity? Would you still have customers if it took you twice as long to deliver the product? Try a little sensitivity analysis, and make the figure 25 percent or 75 percent. Just how dependent on your network is your business?

What about your competition?

A box of books in the warehouse does Wiley no good, nor the bookstores, nor the readers. Information has value based on its possession by someone who needs it. Like every other product, its value is proportionate to the need; the price people are willing to pay depends on the value they place on it as well as on their budget. But information is different from many products in one major respect: exclusivity. If I have a chocolate cupcake, no one else can have that particular chocolate cupcake. But if I have an understanding of the Border Gateway Protocol (BGP), that does not restrict anyone else from having the same knowledge.

Just because the data still sits on your server does not mean that a hacker has not perused it, altered it, and then sold the original specifications to the highest bidder. Imagine that.

Think the Unthinkable

On a day-to-day basis, you think about getting the best performance out of your limited resources. In a cost-competitive environment, you think about squeezing out redundancies, eliminating such a waste of resources. The networking environment has changed, and those redundancies may be your savior in the event of a catastrophe, cyber or physical. You have to change your thinking as well.

In addition to changing your thinking about how to operate your network, you must think about how to destroy it. Especially if you work for a large, visible American corporation, if someone isn't already thinking about destroying your network, it won't be long until the thought strikes a hacker, a hacktivist, or perhaps just a terrorist who knows a hacker.

I confess that the thought of disrupting or destroying a network, just because I can, makes no sense to me. Nonetheless, it must make sense to some because we read of it happening all too often. Of course, the perpetrators allege ignorance of the effects: They didn't know the worm would spread so far and so fast, destroy so many files, even though they programmed it to do just those things. In the meantime, someone else is asking why the backups won't mount.

Plan to Survive

Once you think about the unthinkable, you must figure out how to keep the network delivering the necessary information to the necessary people when and where they need it, despite everything that has (supposedly) happened. You fundamentally have two alternatives to make this happen; I will discuss them when we get to creating your network continuity plan in Chapter 9, "Preparing for Disaster."

You must prepare a business case for the plan in order to get the approval and then the funding. That is the topic of Chapter 11, "The Business Case"; it won't help you if management doesn't agree that you need to undertake an effort on the scale you propose. If your network is like most business networks, it has grown up irrationally: There has been no long-term plan or architectural design to it. Making that kind of network survivable is going to be a significant task. Justifying the work to develop the design—to identify your weaknesses and map out the best way to mend them—will be a small project in and of itself, even before you begin the actual implementation. This book is intended to help you sell the idea to those who can render approval and support your work with funding.

In the process of selling the preliminary work, you will have to explain why some things are needed and what could go wrong. Your audience probably uses the network without any understanding of how it delivers information on demand. The first part of this book, Chapters 2 through 4, is a description for management of your primary threat, cyber-attacks, and why your network, like every other network in the world, is inherently vulnerable to them. You don't want to scare them too badly, but they must become aware of how fragile their information support structure can be when it is inadequately cared for.

Cyber-threats are followed by a discussion of your next most serious threat, a natural disaster, and some questions you must ask about your network and what it does for whom in order to keep delivering that when the world seems to end (locally, at least).

Finally, we turn to terrorist events. The attacks on September 11, 2001, showed us alternative network continuity plans and how well they worked when things went bad in the worst possible way. Unlike the case with many disasters, information about how well or how poorly organizations fared during this catastrophe has been widely available; as noted previously, we will cull a few of the many corporate/government stories and mine them for lessons.

Of course, surviving is not enough. Life does go on, and business will, too. You must include in network continuity how you will recover from your losses and then move forward. Recovery is a part of continuity that is too often assumed to be no problem. You will commence the recovery from an operational condition that is not your normal one, and so you must explicitly consider the recovery and survival phases from the conditions then prevailing to have a true continuity plan. You will find this covered in Chapter 10, "Returning from the Wilderness."

A series of appendices are at the end of the book, with resources you can use in your research as well as bulleted lists of questions you need to ask to determine what your network does for whom, considerations for your network continuity plan, and questions to use in a post mortem; you can at least plan to learn from your own experience. There is also, for those not familiar with discounting, an explanation of how to put all your cost and benefit comparisons in constant dollars so that you can make valid comparisons. Appendix E, "Time Value of Money," expands on the information in Chapter 11, "The Business Case."

Choice versus Chance

"It's choice, not chance, that determines your destiny." We all make choices. I have made two choices that directly resulted in my being alive today to write this book. IBM made a choice to turn away from the business model under which it had developed and been profitable for many years to become far more of a service company than it had been historically; some analysts are not sure how alive IBM would be had it not made the choice it did, when it did.

Your company wishes to survive in turbulent times, but wishes alone are not enough. You will have to convince senior management that the network is critical to the business operating successfully in your current market. Senior management will have to decide how much continuity the business as a whole must have. Based on that, you can determine how much network continuity is required. Once you know that, you can explore your alternatives and cost them out.

The best time to begin was yesterday; the next-best time is now.

CHAPTER 2

Network Threats

One ought never to turn one's back on a threatened danger and try to run away from it. If you do that, you will double the danger. But if you meet it promptly and without flinching, you will reduce the danger by half.

Sir Winston Churchill

One of Milton Friedman's more-trenchant observations was that inflation is always and everywhere a monetary phenomenon. Likewise, network security is always and everywhere a people phenomenon. Why would a person want to damage a network? Why, to be specific, would any person want to damage *your* network?

In one sense, it doesn't matter. Whether you understand the motivations of someone who wishes your network harm, or even those of someone who damages your network inadvertently, is irrelevant. People do damage networks, and the fact that they do is your problem.

Your network is a business asset. In a literal sense, it is an invaluable asset. For one thing, how do you characterize the asset "the network"? What are its physical boundaries, so that you may define the component parts to value and depreciate them? On which schedule do you depreciate these parts? Can you be sure you even know how long some of them have been your property, in order to know where in that depreciation schedule to place them?

Aside from the component parts, do you value the synergy the asset brings in its entirety? For instance, a LAN segment in the Santa Clara office, a LAN segment in the Richardson office, and a LAN segment in the

Raleigh office each have some value, but do they not have a far greater value when they are linked by a WAN so all three offices can interoperate? Have you placed a value on that interoperation, perhaps by the greater productivity of the three offices as they work together? Does that value increase over time (and, if so, have you revalued it lately)?

Before you can protect your network against harm, you must have some idea of what you can afford to spend on that protection; your budget is partly a function of the network's value. It is possible that you could overpay for your protection and have been better off risking damage. Far more likely, however, is that you will underpay for protection, with the result that you will suffer unnecessarily when your network is damaged.

"When" is, indeed, the operative word. To understand why, we need to look at who damages networks and the types of damage they inflict. In some cases, we will see that the perpetrator may or may not even realize damage is being done. Consider this: If someone doesn't realize they are causing damage, what reason do they have not to commit the act that caused that damage? And, of course, if harm was the intent, what can you do to change that intent?

In that sense, if you understand the behaviors that lead to damage, you may be able to prevent that damage. Note that I did not say that you will understand the motive of an attacker, someone deliberately seeking to cause damage to your network. We do not necessarily have to understand the motives of vandals, for instance, to protect our real properties from their actions; if we understand their behaviors, however, we may be able to deflect those behaviors away from our network assets. Unfortunately, vandalism may well be the least of your problems.

Kinds of Attacks

There are a number of ways to categorize attacks on the network (inadvertent damage may take the same form as damage from an attacker; logs will help you decide which it was). We can categorize attacks by their origin (external versus internal) or perhaps by their severity (minor, major, and critical, for instance). An attack on your network, though, is a deliberate act. Someone chose to do this thing. Therefore, I prefer to classify attacks by the skill level of the attacker; this approach tends to correlate with the type, as well as the degree, of damage you suffer.

In this vein, there are three basic forms of attack on your network:

- An attack by immature hands, almost like an apprentice, someone relatively inexperienced (though not necessarily young), who is likely to be probing and exploring as much as attempting to inflict damage.
- A journeyman's attack, which is often a deliberate assault, intended to prevent the network from functioning.
- The master's attack, done with mature hands, whose target may be your network's content rather than its operation.

The last is by far the most dangerous to the business purpose of your network, though the second will draw the greatest attention if it happens.

This order may seem a bit strange at first, but it reflects the level of development of your attacker. Not every attacker goes through these phases per se; they are representative of stages of skill and development in general. Attackers, being human, are individuals and will have quirks or idiosyncrasies, but they will demonstrate certain behaviors at certain levels of development. The form of attack chosen by the relatively inexperienced attacker has certain characteristics; with experience he or she may become more sophisticated and find the challenge of bringing down a network appealing. The truly skilled attacker is the hardest to find—if you even realize he or she was there.

Immature Hands

The relatively inexperienced attacker is likely to show a mixture of sophistication and clumsiness, depending on the resources he or she has been able to find in self-education. To a certain extent, that depends on what resources the attacker has sought. While many people are vaguely aware of the idea of the hacker as a teenaged boy, probably with above-average intelligence but fewer social skills, seeking thrills and trying to impress his friends, this character is as much myth as reality, except in one aspect. We tend to stereotype criminals and, especially those who are wantonly destructive, as being of lesser intelligence.

Could you hack your network, even from the inside?

Do not assume that a hacker is any less intelligent than you are. If nothing else, that will keep your surprises pleasant.

Hackers may be of any age, either gender, and operating from a new computer or an old one, using a dialup account or a broadband connection.

They do have one thing in common: They like exploring other people's systems. In fact, they like it intensely enough to devote hours to the process. They essentially do a great deal of searching in order to find something they consider worth looking at. They begin, perhaps, as *network voyeurs*.

Voyeurs

Think about the popular conception of a voyeur: a bit of a sneak, obtaining gratification from observing other peoples' private activities. It offers the voyeur a sense of power to have knowledge—especially secret knowledge—concerning the private lives of others. The voyeur has defied an invisible boundary in order to make his or her observations.

Networks, too, have boundaries. They, too, have expectations of privacy. Those who violate these may be considered, at the least, voyeurs. One difference between a standard voyeur and a network voyeur is the location from which the observation takes place.

A standard voyeur may observe from outside legal property lines, and the target's only recourse is to block the line of observation. A network voyeur, on the other hand, cannot really see anything of interest until he or she has actually crossed the boundary (more on that in a moment) into your network; the voyeur must observe from within. Real property boundaries are usually marked well enough that there is a reasonable expectation that a competent person will be aware of the boundary's existence and location.

How well defined is your network boundary?

Notice I did not ask how well marked your network boundary is. Whether you're a dog or a real estate tycoon, you cannot mark a boundary until you have defined its location, at least in your own mind. With real estate, of course, the boundary is defined (and therefore markable); based on that definition, administrative control is tendered. With networks, we have something of the opposite: One definition of the network's boundary is where the administrative control changes. This point is often referred to as a *demarc*, short for demarcation point. For traffic to cross the demarc, there must be some agreement between the two network administrations; otherwise, the traffic would simply be discarded, and the link between the two networks would serve no useful purpose.

In addition to that link providing a potential ingress point for a hacker, it costs real money to provide. If the link serves no useful purpose, it is a business expense (and potential liability) that generates no revenue, directly or indirectly. It will not be there; if one is there, as a legacy, perhaps, it should not be there for long.

You do know where your links terminate, don't you?

Bounded versus Unbounded Networks

Another way to consider the issue of network boundaries is to consider the idea of a *bounded network*. A bounded network is one in which all the system's parts are controlled by a unified administration and the network's performance can be completely characterized and controlled. Theoretically, you could identify all of its parts and understand all of its behaviors. Unbounded networks violate at least one (and often more than one) of these principles.

An important point reflected in the concept of bounded versus unbounded networks is that you must analyze the problem as a system. A system is something composed of separable parts, which are often capable of operating independently yet which, operating together, form a unified whole and/or serve a common purpose better than the components individually can. The whole is very much greater than the sum of its parts.

A network, business or personal (in your home, perhaps), which is entirely controlled by one entity and which does not interact in any fashion with any other network, is a bounded network. A bounded network usually has very little business utility. Some business networks, though, are bounded—Research and Development, for instance, may have a bounded network.

A bounded network is not inherently safe from attack, of course. Electrons, however perverse they may seem at times, do not go someplace unbidden. An attack is a human-initiated event; attacking a truly bounded network simply must be done by someone with physical access. That someone could be an imposter or an otherwise unauthorized person, but it could very easily be an authorized person doing something he or she was not supposed to do.

Most networks, including the overwhelming majority of business networks, are unbounded, if for no other reason than, at least at one point, they connect to another network outside the company's administration (such as the Internet). While bounded networks are not inherently safe, unbounded networks (to be delicate) are less safe. Unbounded networks suffer from several problems: They not only face distributed administrative control, but there is no central authority that can coordinate among the administrative entities. Likewise, instead of being able to know all the parts, there is limited visibility beyond the local administrative unit; there are parts to the system you may know nothing about. There simply is no complete set of information about an unbounded network.

A corporate network, even one composed of many LANs, is bounded until it has a single connection (authorized or not) to a network outside its administrative control, such as an extranet or the Internet. As a result, it is common for a system to be composed of both bounded and unbounded networks. For instance, a firm may have three regional domains, all bounded networks, plus an extranet connection from the headquarters domain, such as that shown in Figure 2.1.

Because you do not administer the extranet, except (perhaps) jointly with the other extranet members, once the connection to it is established, your network (all four domains) has become part of an unbounded system.

Some business networks became part of an unbounded system when the CEO wanted—and got, of course—a modem on his phone line to check his AOL account. There is nothing inherently wrong with having or accessing an AOL account, but accessing it from anywhere except within AOL's own network means your network has instantly become unbounded. This was the essence of the security issue for former CIA Director John Deutch.

Your company's network is part of an unbounded system.

Defining a network's boundaries is probably simplest by considering everything you administer—everything inside all of your demarcs—to be your network. Every demarc is on a link and is therefore an entry point for incoming traffic, or an ingress. Of course, it is also an egress, or exit point, for traffic as well. Like a traffic intersection for surface transportation, both directions must be considered. After all, if the network voyeur can gain entry but never realizes it because the flow of information back to her is suppressed, she will not find your network interesting and will probably move on.

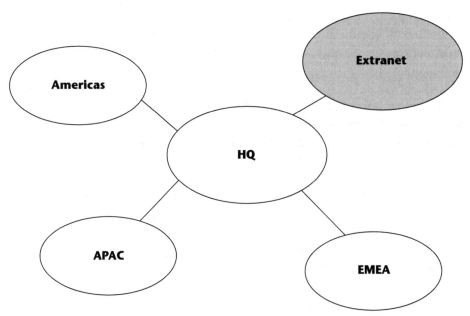

Figure 2.1 An unbounded system.

In thinking about the threat to your network from the voyeur, you should reconsider how standard voyeurs are kept at bay. We ensure that property boundaries are both well defined and well known. To preserve our personal privacy from those who would pry from a distance (using optical aids such as telescopes or high-powered binoculars, cameras with telephoto lenses, etc.), we create internal barriers to block the reflected light that transmits our image.

The same principles apply to protecting network privacy from network voyeurs. You establish boundaries and make them known so that there can be no doubt that someone is trespassing if the boundaries are crossed (more on that in a moment). You also obscure the view of anyone who attempts to pry anyway.

Why the emphasis on making your boundaries known? Especially with the explosion in network access made possible by the similarly explosive growth in the Internet, unboundedness has become the norm. It is as though an entire neighborhood were suddenly, vaguely, one amorphous property within which anyone—resident or not—was free to poke around at will. Because that description is apt in many places, you must ensure that it cannot be applied to your network by extension; you must differentiate your network from the freely accessible internetworks of the world. To protect your property rights, you must demonstrate that trespassers should have known they were violating your boundary.

That knowledge—or, more accurately, the lack of that knowledge—is part of the problem. While hackers are often curious and derive great pleasure from solving challenges, they do not usually realize they are trespassing on someone else's property. Their crime (and trespass is a crime in most, if not all, societies) is done from a distance; they see no impact on your network or your business. Add this action-at-a-distance to the norm of unboundedness, and the result is predictable: In their own eyes, they have done nothing wrong. They were just looking around, through a wide-open door or window. If you didn't want them to see, you should have closed the door. It's your fault.

They may even claim that they are doing you a favor. Like a cyber-patrol Neighborhood Watch, they have passed through and checked your doors and windows. You left them unlocked! You should know better. Thank goodness you finally noticed me standing inside the (closed, though unlocked) door—anyone might have got in, and think what damage that sort of person might have done. Interestingly, you may even hear this argument from someone who, to continue the door analogy, had to pry away the framing around the door in order to lift it out of the way. If the hackers could do that, of course, an evildoer might as well, so (again) you should be grateful for their showing you how flawed your home's construction really was. You'd better fix that, and soon.

In one respect, it is your fault, even by a common-sense definition as opposed to such a self-serving one. If you have not made clear where your property begins, you cannot reasonably expect someone to honor the boundary. And the more challenging the entry, the more talented the person who finally trespasses. This is a mixed blessing: You will hold off the least experienced, most immature hands (the majority), but those who do gain entry are more dangerous. They may even be able to inflict serious damage (our next major topic).

Further, these are systems, not just isolated machines. One unknown ingress, one unpatched flaw in the operating system, even an unneeded process available (and some have design flaws) can leave you open to observation. Because some hackers do become skilled enough to dismantle the architecture in order to get in (sometimes for no other reason than because it became a challenge), you must expect to have voyeurs. In that event, you bolster your legal case by ensuring they cannot claim ignorance of their trespass.

Boundary marking is best done right at the ingress; normally, this is a router. Ports attached to external links should be filtered against unwanted traffic, and unwanted or unexpected traffic should not be allowed in. You don't invite the voyeur into your home or office to have a better look, after all. You can also make clear at the router port (through a banner or Message of the Day) that this is the property of the XYZ Company, Inc., and non-business access is not allowed. You may even wish to warn that you will prosecute trespassers; make that choice after consulting with your legal advisors, as laws vary from one physical location to another. Jurisdiction will also be a factor because the person who committed the unallowed access is most likely not local. Because the law is evolving in this area, you should both consult with your legal advisors and periodically review the matter.

Testers

Different cultures have different rites of passage. Almost all require the initiate to meet a test; this could be a test of physical courage or adaptability to stress. It could also be a demonstration of knowledge or capability. A knight had to earn his spurs; the doctoral candidate must defend their dissertation. You must succeed as a manager to be considered for promotion to senior manager. To graduate to the next level of our chosen path, we must demonstrate that we deserve to be there.

Not all apprentices are content to remain voyeurs. They have learned the methods of voyeurism from a culture; being human, some wish to advance in their standing in that culture. To demonstrate their skill, their work has to be visible to others. Screen captures of someone's router interface command line can be faked, but if the news reports that XYZ Corp. discovered a planted file—nothing damaging, say, just a birthday greeting to the CEO— suddenly appearing as the Message of the Day (MOTD) on its routers, then the candidate has demonstrated that he or she was there.

The tester is in a transition phase from passive observation to active disruption. He or she is testing their own skills in active engagement with you as well as your network's security. The intensity of the hacker's emotional reward will grow as they succeed in meeting more difficult challenges.

Deliberate Attackers

With more skill in the hacker's hands, more experience in his or her toolkit, the hacker may not be content to merely browse and perhaps prod you from time to time. You haven't really reacted so far, after all; why not find out just how much it takes to get your attention?

HACKERS VERSUS CRACKERS

Hacker, cracker, who cares what you call them? ("What's in a name? That which we call a rose/By any other name would smell as sweet;/So Romeo would, were he not Romeo call'd...")

That was the rub, in fair, fictitious Verona; names did matter because they identified not only what you were called, but who you were. They had baggage.

Names still have baggage. A hacker was supposedly a white hat, a Good Guy, while a cracker is allied with the forces of evil, hell-bent (so to speak) on sowing doom and destruction wherever he or she telnets. Naturally, while such bright-line distinctions make good theater, theater is not life. If it were, how would it sell? Life, in the main, is interesting only to those inside it; theater must interest you from the outside and draw you in—it must be better than life, larger than life, and its characters more brightly defined than those in real life.

Is there a bright-line distinction between a hacker and a cracker? No. Do some people claim there is? Of course. Those with a vested interest in being identified as one or the other will make the differences between them as distinct as possible. In fact, the lines are quite blurred, by both time and behavior.

(continues)

> **HACKERS VERSUS CRACKERS *(Continued)***
>
> To consider time first, *hacker* originally meant someone who took things apart all the way down to their base components, just to understand how things went together and worked. They hacked at the problem until they understood it in all respects.
>
> Hacker was originally not just a compliment, it was a supreme compliment. Hackers often understood a system better than its designers. To an extent, that reflects the difference between theory and practice or between the scientist and the engineer. The hacker considered the system as it operated in the real world and found aspects of its behavior that the design team never expected to emerge.
>
> Hackers notified designers of the flaws they found, sometimes in research papers, sometimes by email or posting to topical newsgroups. True hackers took no advantage of the flaws they found; they warned others of the existence of those flaws so all could take remedial action.
>
> This changed as networking connected more people, not all of whom had noble intent. Some wanted to be respected as hackers but had not reached the level of understanding required; they could, however, capitalize on the flaws identified and demonstrate that fixes had not been applied. They went the same places and showed the same things—they must be hackers, too.
>
> The original distinction, then, was between the hacker, who examined the entrails of the system beast in order to learn about the beast, and the cracker, who opened the beast not to examine the entrails and further everyone's knowledge, but merely because they could ... and, while there, they might as well have a bite or two. No one could ever say the cracker wasn't there if they left his or her mark behind.
>
> With such differences, how could the lines be blurred in either time or behavior? Over time, journalists have tended to call all who broke down a system hackers, whatever their intent or subsequent actions. As for behavior, even the original, good-guy hackers trespassed, did they not? Except in a network laboratory, even the good guys entered networks without permission. They were unbounded systems, but true hackers knew, if anyone did, where they were. And they rarely asked permission; they simply entered and then warned that others could, as well.
>
> Intent may affect the degree of the crime, but it does not change the fact of the crime.
>
> And even if a good-guy hacker snooped through your network, how can you be sure he or she only looked, but did not touch?

Even though the voyeur has trespassed and violated your property boundaries (roughly equivalent to breaking and entering, but not yet stealing anything), he or she didn't change anything, so your network continued to operate as before. When the hacker transitions to tester and then deliberate attacker, the problem escalates. Like crossing a tipping point, the nature of your problem is irrevocably different. Now, everything about your network must be suspect. A major business asset must be presumed to be flawed; you must find the flaws (and you must assume there are more, perhaps many more, than one), and you must repair them, but can you ever really trust the character of the network again?

Some attackers are barely beyond the voyeur in skill level. Hackers often described how they accomplished some entry or the consequences of an action on a process. Other hackers then created scripts to automate the steps involved to achieve the same things. They published those scripts where even less skilled (and possibly less scrupulous) individuals could find them.

The less skilled players often demonstrate that they deserve the dismissive tag *script kiddies*. They download and run the scripts to gain access, and then quickly demonstrate that they have no idea what they are doing (as an example, they try to use DOS commands on a UNIX system, even though the two operating systems have different prompts as well as appearances at login). Their danger to your system lies less in the damage they can inflict on their own than that which they can cause using tools created by others. Even an ancient charioteer could mow down quite a few soldiers if handed a machine gun. The first time something went wrong (such as the mechanism jamming due to overheating from continuous fire), he would be at a loss, but his prior targets would be dead, nonetheless.

Unfortunately, the tools are becoming both more sophisticated and more widely available. Simultaneously, networks are becoming more exposed, a subject we will turn to in Chapter 4, "Murphy's Revenge." The tools available now include scripts to search out unwitting accomplices in the form of computers attached to a network and left on, unsupervised. An attack from one source is much easier to counter than one from many places at once (an ancient Chinese torture was known as the "Death from a Thousand Cuts").

Assume, however, that you are not the target of a script kiddy, or even several of them. Assume, instead, that your network is in the crosshairs of one who has passed beyond that level. He or she may not possess the expertise of a true hacker, but he or she knows enough about systems to get in without help and that when they change *this*, the outcome will be *that*. This person can cause your business problems, even if they do nothing to disrupt your network. Perhaps he or she merely makes your company look foolish.

I assume you would prefer to avoid that.

If you are wondering why anyone would want to make your company look foolish, there is a (relatively) new player in the hacking game: the *hacktivist*. Using the skills of hacking to promote an activist message means that your company's electronic presence is now vulnerable to disruption for social agenda reasons in addition to all the business and legal issues you have previously contended with.

Whether done for feelings of personal power or to promote a cause, network attacks are an escalation from snooping and browsing by a voyeur. The latter left things untouched (can you really be sure?), while the former seeks to change your system for their benefit. The attacker may have created private backdoors, so that they can come back at will and do whatever he or she wishes to your property. If that thought makes you uncomfortable, good; you should be.

Voyeurs to testers to attackers to true hackers is a continuum; despite informal rites of passage, there are no graduation certificates from one stage to another. Their potential to cause damage is likewise a continuum, with overlapping zones (see Figure 2.2).

At some point in the range of the attacker, we pass from the realm of immature (inexperienced) hands to that of mature (experienced) hands. The type of damage caused is not another continuum, but the sophistication of the damage will reflect the sophistication of the perpetrator. Because the voyeur has little actual technical skill, his or her damage tends to be minor (relatively speaking) and clumsily enough done that it is readily detected. When we reach the level of the expert, the damage is much more expensive—if you even detect it.

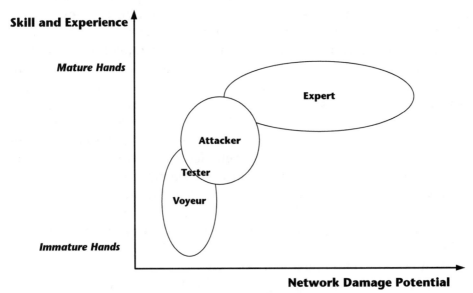

Figure 2.2 Damage potentials.

In the middle range, the operating arena of what we've called the attacker, the damage is more serious than the fumbling attempts of the voyeur and more visible than that of the expert. Because of this combination, this is where you may be tempted to spend most of your company's time and effort. That may or may not be wise; it is your decision to make, based on your business's operating characteristics and legal exposures.

What damage is likely to come from an attacker? More importantly, what is the effect of an attack on your network? *Attack* brings with it connotations of combat, if not outright war. Hacktivists, of course, do seek confrontation at the minimum, and some indeed see themselves locked in mortal combat with the forces of *<insert cause/enemy here>*, as epitomized by *<insert your business here>*.

To evaluate the damage attackers may cause, think back a moment to our discussion in Chapter 1, "Introduction." Why do you have a network? What is its business function? A simple description is that the network is your company's nervous system. It is where the information is stored and accessed; it is how commands are transmitted and feedback returns to the brain. It is not the brain per se; corporate management is that. But the network extends the reach of all employees, not just headquarters; the business can reach more people, do more things, and do them more effectively and economically because of the information distribution possible with your network.

What is the effect of an attack on a human nervous system? Paralysis. Uncontrolled motion or speech. Impaired cognitive functions due to inadequate or incorrect input or impaired processing of correct information. Certainly reduced effectiveness. Possibly even death.

Make the parallel to business, and the prospect should concern you. Panic certainly won't help, though when an attacker runs through your network faster than IS can track him or her, panic might well develop. Instead, you must take the measure of the possibilities in advance, determine the cost of prevention and/or mitigation, and decide how much of each you can afford.

A network attack is similar in its effect to the military blitzkrieg attack. The goal is not to destroy the enemy force, but to render it incapable of countering what you wish to do. Blitzkrieg developed from the convergence of two factors: new technology interacting with old tactics. Beginning with the American Civil War (1861-65), through a series of wars until the carnage of World War I (1914-18) tactics did not change despite the introduction of completely new technologies and their continued refinement as weapons.

The blocks of troops advancing en masse, so effective in the Napoleonic Wars, were decimated by accurate rifle fire in the Civil War; the addition of Gatling guns late in the war (the first effective machine gun) only made it worse. Incredibly, over 50 years (and multiple wars) later, ranks of infantry were still being sent forward across fields of fire from better rifles and even more improved machine guns. In Europe, the carnage was such that an entire generation of young men was lost. Effectively, that is accurate: Those who survived were forever changed. By the beginning of World War II, they were the leadership, both military and political.

Their reactions to continued (and increasing) tensions took one of two forms. One reaction was to do whatever it took to be sure such a conflict did not happen again. This group was epitomized by the British Prime Minister Neville Chamberlain, famous by obtaining "Peace in our time," a peace that lasted less than a year. It is important to remember that these men acted rationally according to their view of conditions; though their actions look foolish in retrospect, many, if not most, of their electorate approved of what they did.

The other reaction was based on the underlying assumptions that conflict and war would happen again because they were as old as mankind and that technological change mattered more than political change. The epitome of this school of thought was Heinz Guderian. Because conflict would come, what had to be avoided was the direct confrontation of your flesh with their machines. The way to do that was to keep the enemy from bringing his force to bear on you; strike fast, strike hard and deep into his territory, and keep him from thinking—make him react to you, do not ever have to react to him. In essence, blitzkrieg prevented the enemy from thinking about what he was doing, by disrupting communications to and from the head(quarters), and by making things change so fast that any communications received either way were irrelevant. As the enemy fell further and further behind the power curve, he might panic, seek to withdraw and salvage what little was left, surrender on the spot, or do some combination of these.

To take on a corporation once meant that massive manpower had to be involved; whether letter-writing campaigns from consumers or organized labor actions, it took the continuing, concerted action of many people to gain the corporation's attention and interest. Forcing change might take a multiyear campaign.

Network attacks need none of that. They disrupt your ability to use your information, to think. During the attack, you fall further behind the power curve. You know things within your network have been changed. Yet even while your best people are trying to determine exactly what has changed, the nature and extent of the damage continue to grow. You don't know where the attacker has been, what damage he or she has done, where he or she is now—even if the attacker is present within your boundaries. Management has been known to panic, making the job of IS personnel countering the attack even harder. The attacker has not engaged you in a frontal assault, wearing all your people down and destroying your equipment. In fact most of your company is intact (including your hardware, software, and data); you simply cannot act effectively.

Network attacks are very like a blitzkrieg.

Depending on the nature of the damage, communications—the business reason for the network as well it its own sine qua non—may be rerouted, intermittent, completely cut off in some places but not others, and so on. The company's brain is still functioning, but the nervous system cannot reliably deliver its messages to the rest of the organism or report conditions from the organism as well as the outside world to the brain. Your company may seem stunned, or it may convulse; parts may struggle and manage to perform adequately, but at great cost. And once the attack is contained, then eliminated, the repair and restoration process will take longer and cost more than expected, if for no other reason than the fact that you will discover further damage as things settle down and you are able to take a proper assessment of your system.

This is the kind of event that most people imagine, in insufficient detail, when they talk about a network attack. The number of occurrences of such events is increasing. The Computer Security Institute has conducted a survey every year, for the past six years, in conjunction with the Federal Bureau of Investigation's Northern California office. The survey results have been available as a free download at www.gocsi.com. The survey respondents are the computer security professionals at many large corporations and government agencies in the United States. The trends they report are both encouraging and discouraging.

On the encouraging side, the percentage of respondents whose companies were taking stronger protective measures continued its trend upward. Also encouraging is that a larger percentage of those attacked reported the incident(s) to law enforcement authorities. Another item that is somewhat encouraging is that the response "don't know" occurred less frequently in the most recent survey than it had in the past. These large organizations are being more proactive about securing their networks.

That is good because they are being attacked even more frequently than they were in the past.

According to several measures, such as the frequency of attacks and the dollar cost of those attacks, the problem is getting worse, not better, despite the stronger defensive measures being taken. The point of origin of network attacks was once roughly 4:1 inside the network (80 percent from inside the network versus only 20 percent from outside), though that did not necessarily mean they came from an employee. Contractors or those with inside physical access could also have been responsible.

The old 80-20 rule no longer applies. In fact, it hasn't applied for four years; since the survey began breaking out attack origin, more attacks on these networks have come from sources external to the network than internal. How well-defined *is* your network boundary? What *is* on the other end of your ingress links? One answer to the latter is the Internet.

I do not suggest that you should not be connected to the Internet; it is entirely too useful a resource to avoid as though it were somehow contaminated with a network plague. Among other things, your employees probably have come to expect Internet access; if you cut it off, productivity will likely decline for direct reasons (it becomes more difficult to access certain information) as well as indirect ones (morale).

In Chapter 3, "Tactics of Mistake," I will examine the forms an attack may take; for now, accept that if you are connected to the outside universe, you have an unbounded system. Because you do not entirely control an unbounded system, you cannot prevent attacks. Your options lie entirely in the realm of how you handle them.

Mature Hands

One interesting aspect of the CSI survey was the cost reported by those who were able to estimate it (and willing to share that information). Overwhelmingly, the greatest losses suffered were due not to the attacks I have been describing, but rather to those requiring more skill—those of mature hands. Theft of proprietary information and financial fraud were far and away the most expensive form of network attack, with viruses coming in a distant third. Data theft also occurred more frequently, so that the total damage reported from it was two-thirds greater than that for financial fraud. Remember this when you estimate how much you can afford to spend on protection.

Some of the theft may have occurred whether or not you use a network. *Social engineering* is the term used for getting information from people they really shouldn't have told you. Examples abound, from the story of a firm, well ahead of its competitors, whose employees supported all the information requests from a consultant supposedly brought in by the (vacationing) CEO, to an example demonstrated live by a security consultant to an incredulous audience, where he talked a well-known bank's help desk into giving him a series of PINs over the phone.

We train our people to help each other, to work as teams, to make sure we get the job done. The social engineer simply exploits those characteristics of your human network, but social engineering, while still being done, of course, is no longer the only way to obtain such information. The master hacker accomplishes the same thing by exploiting the relevant characteristics of your data network. Both thieves are masters of their craft; neither result is what we might call good for your business.

There are essentially four varieties of attack from the master's hands: industrial espionage, fraud/theft, record alteration, and extortion. The last may seem a bit out of place, but its position within this realm should become clearer as we go on. Each of these has historically been done via social engineering, but it is no longer necessary (in many cases) to go to that expense and risk; your network exists to deliver information to and from storage. Master hackers can insinuate themselves into your network and do damage from the comfort and safety of their homes, wherever in the world that might be. A master hacker also has one distinct advantage over the social engineer: No one ever sees the attacker; if you detect what they have done, you likely have no idea who they really are, what he or she looks like, or even where the attack originated.

Industrial Espionage

It is easy to assume that industrial espionage is not a threat to your business; children also find it easy to believe that money was left under their pillow by the Tooth Fairy. As an adult, you know better about the second belief. As a thoughtful manager, you ought to know better about the first. If nothing else, be aware of the relative costs of information development versus theft.

As an example, assume you are the project manager of a major new software product. It could be a major release substantially improving an existing product, or it could be an entirely new product. You have a small, skunk-works kind of team doing the development, perhaps as few as 14 technical and support staff. You have worked 18 months on this project, and it will knock the socks off the market; the competition has nothing like it and you know from comments made to the industry press that they cannot get their version, plus or minus a few features, to market any sooner than a year after you debut. Whoever delivers first will have a very strong market position, a fair reward for your identifying the problem and addressing it first.

What has been your cost to date? At an average fully loaded labor rate of only $150/hour (a low estimate for this kind of talent), labor cost alone has come to $6,552,000. Add in the proportionate overhead cost charged against this project, plus your time and overhead, and the time and overhead of the senior managers who have periodically reviewed the project at its internal milestones, and a total project cost of $10 million is not unreasonable.

First-year revenues are expected to be at least $8 million, a rapid payoff that will only get better as sales grow, even though your competitors will begin shipping product at the end of that year. By being in the market first, you will have set the standard and should dominate sales for at least another three years after that. Altogether, a good financial picture presents itself to you and your firm.

The picture is not so pleasant for your competitors, however, for they are on the other side of the problem. Not only will they lose the first-mover advantage, their costs are somewhat higher as they try to push the pace and narrow the time gap between your product introduction and theirs. They still face a great cost to be, at best, in second place.

On the other hand, they could spend a relative pittance to acquire your information, make some cosmetic changes, and tie, or even beat, you in the market introduction. They would at least share that first-year revenue stream, at less development cost than you, and they would co-own the market (or possibly totally replace you as the standard-setter).

How much of a pittance? The KGB paid the friends of Markus Hess approximately $18,000 and some cocaine in 1987 to steal a plenitude of secrets, such as manufacturing techniques for gallium arsenide chips and the code for computer-aided design and manufacturing. Hess, who did much of the actual stealing, got a portion of the money and no cocaine. Hess had become a sophisticated hacker for the challenge; it took very little money to reward him for doing what he was doing anyway.

Suppose it cost you (a competitor) as much as $100,000 for the information. And suppose it saved you only half of the total development cost. Spend $100,000 to save $5 million and preserve a position in the market for future sales: Is this choice difficult?

Morally, it should be; financially, it is a no-brainer.

That, in a nutshell, is the case for industrial espionage. There is no case in which you can consider your firm immune. The term *netspionage* is coming to be used for industrial espionage conducted via a network penetration, and it is becoming recognized as a serious problem as businesses conduct more and more of their information exchange over networks that are too easily penetrated. Whatever you spend to develop your product, it is worth up to that amount to steal it. The thief gets the information without the time investment needed to develop it on his or her own, but even properly discounting all outflows to their present value, stealing information is a bargain.

Only 6 percent of the respondents to the CSI/FBI survey were able to quantify their losses due to theft of proprietary information, but those losses totaled over $151 million. One respondent's loss was estimated to be approximately $50 million. Even discounting that extreme instance, the average loss was in excess of $3 million. In other words, these firms could have spent, on average, up to $3 million each on network security and been better off (because their information would still be proprietary).

How much did it cost you to develop your information? How much would you have to spend to regain that position in the market? These answers should help you answer a third question: How much do you spend to secure it?

Fraud/Theft

Unfortunately, stealing information is not the only reason to sneak into and out of a network; stealing money will do as well. In 1994, a Russian hacker (of sorts) stole approximately $10 million from Citibank by manipulating its online transaction system that used dialup connections. It wasn't a network attack in terms of manipulating protocols or UNIX processes because what he actually manipulated was the telephone system over which the online banking was conducted. None of that, of course, changes the fact that he stole the money electronically.

Remember the idea of unbounded networks? Your network is unbounded in another way if it interacts with a network of another type, such as having dialup access that can conduct any kind of transaction. The transaction, by the way, need not be theft; to fit in this category it could be fraud, perhaps inducing you to pay for information or a service you never received. The fraud could be accomplished by simply posting a verification of completed work to one of your functions that purchases outside goods. Unless they have a careful auditing function built in, it could easily be passed to accounts payable. You may have protected the network accesses from the Internet and your extranets, but what about interactions with the telephone network?

Of course, there is also the theft of information that can be used to obtain money virtually immediately. If you conduct any kind of business in which credit card numbers are transmitted electronically into your network, you are an interesting target. The problem is not eavesdropping during the actual transfer of the credit card number, even though that is what most consumers worry about (hence, we have browsers that show a lock or a key to indicate a secured transaction). The problem is that you store those numbers so that the consumer won't have to type the 16 digits, plus expiration date, into a window and worry about the transmission all over again.

Your storage is the target.

Why waste time and bandwidth, not to mention the risk of exposure, hanging around on a connection to collect one or two credit card numbers when there is a file of hundreds, even thousands of them? Explore the network—quietly—and locate that server. Is it protected from access by a trusted host on your network? Quite possibly not. A true hacker can disguise his identity to the network (called spoofing) and appear to be a legitimate host, one trusted with that very access.

Another gold mine is available and, fortunately for the thief, not always as well protected. You make backups of all important files ... like the credit card numbers file. Prudently, you store that information at another physical location. Is that location as well monitored and well protected as your primary location? Is the data stream (most likely sent at a regular, *predictable* time) between the two locations encrypted? By what algorithm, at what strength? Is the key secure?

If the file transfer is intercepted and copied en route, it can be saved until the encryption is broken, at which time the file may as well be plain text, for it surely will be soon. In 1997, the FBI intercepted an intended transfer of approximately 100,000 credit card numbers for cash. The hacker (and he wasn't a terribly skilled one, more of a journeyman than a master—which contributed to his being caught) had copied files of credit card numbers from the customer databases of online businesses. Even if encrypted, for the value to be gained from such a file, the effort to break the encryption is worthwhile. You can make the same calculation as for the netspionage example, but this time use an average credit limit of $8,000 per card, 100,000 cards, and a cost for decryption of the entire set of perhaps $25,000.

Not all encryption algorithms are what their marketing makes them out to be (this should hardly be a surprise to any consumer). Through improper or short-cut processes, their actual strength *as employed* is sometimes less than half what it could be. As an example, the 64-bit key used for European GSM phones can be broken in the time it takes to force a well-constructed 30-bit key, which is a few hours on a reasonably strong desktop computer. Many encryption algorithms are based on a password or passphrase; because people tend to use what they can remember, the basis of the key is far from random, weakening the encryption by weakening its foundation. The cost of $25,000 was exceedingly generous; once a criminal operation was going, the cost would likely fall to just hundreds of dollars.

Willie Sutton robbed banks because that was where the money was. The money now moves through networks; the smart robbers are already there with it.

Record Alteration

A major carrier has made a number of advertisements focusing on data being handed off from one provider to another as it travels across networks, implying that this is less than desirable, perhaps from a safety or integrity standpoint. Data integrity is crucial; if you cannot rely on the soft copy of your data, the network offers you no speed advantage at all. You must revert to keeping hard copies, retrieving them and faxing or mailing them, with all the attendant delay.

Further, hard copies cannot be searched, cross-indexed, or combined—all advantages of relational databases. The same action done one week later, though, must yield the same result (provided it was done with the same algorithm). The data cannot be allowed to change except by your intention. When you think about protecting data integrity, you need to consider *why* data would be altered, *what* can be altered, and *how* it can be altered. This is similar to the hoary crime-solving rule of motive, means, and opportunity; because data corruption is similar, at least in principle, to forgery, that is appropriate.

Why?

The short answer to *Why?* is *money*. *The Sting* was a wildly popular movie glamorizing an elaborate scheme to doctor the information used by organized crime to make money from betting on horse racing. Organized crime delayed the information long enough to ensure their bets won; the scheme altered the information on which the criminals placed their bets in order for our heroes (good hackers, as it were) to turn the scheme to their own advantage.

Likewise, consider a contractual relationship between a vendor and your firm. The contract contains performance guarantees, premiums for exceeding requirements, and penalties for failure to meet minimum requirements. The vendor, of course, would prefer to receive the largest possible payment while you would prefer to tender the smallest possible. The size of the payment depends on information—delivery dates, performance characterization, payment characterization (when due versus when paid), and so on.

Both parties have a financial interest in the information appearing a certain way; of course, those interests oppose each other. If you keep the information, the vendor must be able to validate your claims of payment terms (or be able to take your word for it—possible on a $50 purchase order, but indefensible on a $50 million contract). Likewise, if the vendor simply presents you with an invoice, you must be able to believe it or validate it.

The integrity of the information must be unquestioned; the alternative is to meet in civil court and hope that you recoup your expenses as well as receive the terms for which you contracted.

What?

Any information relevant to an issue can be altered if the data can be accessed. Email headers are often altered by spammers in order to prevent detection or backtracing. An electronically tendered invoice may have the quantity owed modified, by anything from a trivial amount (banks earn a great deal of their income from trivial amounts per transaction, over millions of transactions) to one large enough to cause the entire invoice to be questioned. Delay could be the goal here if the hacker's purpose is to disrupt cash flow by tipping payments into a different period or causing sufficient delay to eliminate eligibility for a discount.

The same invoice could have a small alteration in the name of the payee or the address (physical or electronic) to which payment is to be sent. The invoice requests payment for services rendered; the terms of service in the original contract could be altered if the contract is stored in soft copy accessible to the hacker. When the invoice is questioned, contract lookup validates it.

Performance requirements, or the measurement against them, could be altered, changing the size and timing of the payment to be made. The possibilities are bounded only by the imagination of the criminal and the data you have in storage.

How?

Data is simply a series of binary digits (bits). Change the right bits, and you have changed the information. A change can be made to occur during data transfer or in storage. Change during data transfer is the more difficult operation, although access may be easier; the opposite is true for data in storage.

As data is flowing over the wire (or fiber), it is easier to manipulate because many devices have legitimate access to the same circuit. One of those devices can be compromised, allowing an illegitimate party to gain access in the guise of a legitimate one. Isolating the particular data stream and manipulating it in real time, however, is a challenge (though it is not impossible). An easier task is copying the data, performing the manipulation offline, and then replacing the data in the stream. The corrupted data will arrive later, to be sure; that in itself is a good reason to have accurate network time synchronization and auditing (which could compare transmittal and arrival times of information).

Alternatively, if the hacker can access the storage location itself, there is no need to worry about capturing the initial data transfer. He or she can modify the appropriate bits at a time of his or her convenience instead of lurking, with the constant chance of discovery. Of course, data storage locations are often better protected than the data transfer media because much of the media's content requires no real protection at all (such as an email request for a vacation date). Therefore, it is more difficult to access the data in storage; however, because it is stored in a structure based on the program that created it, a copy of the same program can easily modify it. Even easier for the skilled hacker is to simply insert the changes into the raw data by knowing the location in the structure where the relevant information is located.

Extortion

That brings us to the fourth type of attack by a master: the bluff. Unfortunately, of course, you don't know with certainty that it is a bluff.

When you are notified that a hacker has threatened to attack your system, you may be tempted to play the odds (as you see them) and ignore it. What reason do you really have to believe that this is not some overgrown little boy crying wolf?

Typically, the extortionist offers proof, in the form of some data that they show you that they have stolen and/or accessed and/or altered. The extortionist recommends that you verify that he or she has done what they say they can do again. From this, you know two things: one, they feel they are dealing from a strong hand because they are giving you information, and two, they feel very confident that they can do it again, even with your knowing what they have just given you.

If the extortionist's assumptions are correct, this can be a bluff because you must respond. Whether you choose to accommodate the extortion is a business and ethical decision. You must assume that the extortionist has ensured that they can get back in. You must assume that they are capable of delivering on the threat. On the other hand, as Churchill once reminded Britons about appeasement, "Once you pay the Danegeld, you never get rid of the Dane."

Externalities

Externalities, to an economist (and most business people suffered through at least two terms of economics), are aspects that are not captured in the price mechanism; positive externalities are received benefits not paid for, while negative externalities are incurred costs not compensated for.

In the CSI/FBI survey, the number of companies reporting internal-origin attacks and the number reporting external-origin attacks were very similar, though the number of attacks themselves was heavily weighted to external origin. In an internal attack, you're getting something from your people for which you are not being compensated (negative externality); they are certainly getting something for which they did not pay with the work you employed them to do (positive externality).

As we will see in the next two chapters, it is much easier to secure your network from external attacks than from internal ones. Because the internal attacker has the benefit of a working knowledge of your network as well as a point of origin already inside the first line of defense, the internal attacker is in a position to do far worse (and far more expensive) damage than the hacker nosing in from outside.

Externalities are one of the trickiest problems to resolve in economics—the gainer doesn't want to lose the freebie, and the loser feels cheated. It becomes very difficult to adjust the price to incorporate all the effects of the transaction and still keep both parties willing to participate. Your network's security is an ongoing transaction between you and the people you have trusted enough to have access to it from the inside. Our next step is to understand the details of how you can be hurt.

Tactics of Mistake

Beware of Greeks bearing gifts.
Cassandra (allegedly)

It has been said that lieutenants and captains study tactics, majors and colonels study strategy, but generals study logistics. To an extent, that is so because of the relationships among them; tactics are employed in support of a strategy, but without the proper logistics, the strategy will fail. It also reflects the capabilities for understanding of the officers involved. Younger, less experienced officers study the smaller pieces; middle-aged (in military terms) officers pull those together, tempered by their own tactical experiences; mature leaders create the operating environment in which the others can succeed.

This is a chapter about the tactics your attackers will use. As with the building blocks of military thought, once we understand the tactics, we can address strategy by looking at how and when the enemy is likely to use certain tactics; we can then position ourselves to best counter the enemy. This does not—repeat *not*—mean that you can position yourself never to be attacked. The only network that will not be attacked is the completely bounded one, whose power switches are collectively in the "off" position, and that no human can physically access. And even those conditions can be gotten around if the attacker has the right incentive.

Because you will be attacked (actually, you have almost certainly already been attacked, whether you realize it or not), it helps to recognize the tactics being run against you. Further, I do not recommend counterattacking and attempting to destroy your attacker. For one thing, the attacker is not necessarily who you think he or she is. For another, as early as Sun Tzu, over 2,500 years ago, strategists have counseled to leave the enemy a path to retreat in order to encourage him to do just that (rather than fight to the death and take more of your people with him). In defending your network, by all means do everything you need, but do no more. This is not a game, and it has no fixed endpoint after which we all shake hands and go home. You must be able to outlast your attacker, and that takes logistics.

You may expect a variety of attacks, ranging from not directly damaging to devastating. The least damaging attacks are the probes—reconnaissance, to continue the military analogy. Any of the remaining broad types can severely degrade your network's performance or even take it down completely. They are viruses and worms (similar in principle but different in completeness as well as execution), denial of service and distributed denial of service attacks, and Trojan horses.

Before we can explain what those tactics do, you may need a refresher (or a primer) on how packets move through a network. If you work with IP packets routinely, this is old hat; skim or skip to the next section as desired. If you haven't worked with IP for a while or have used IP networks but never needed to know how the information moved around in them, this will introduce the principles on which that movement is based. These pieces are what attackers use against you.

TCP/IP

TCP/IP (Transmission Control Protocol/Internet Protocol) was developed for the U.S. Department of Defense in the early 1970s. The goal was to have a flexible, survivable means of routing electronic information. Computer networks were, if not in their infancy, mere toddlers at that point; networks in general and TCP/IP as a networking protocol suite have evolved over the intervening three decades (three decades, of course, is multiple generations in Internet time).

TCP/IP focuses its attention on the interface between the local host (a particular machine, such as a client, server, network printer, or router) and the network to which it speaks. You may hear reference to a "layer" by a number; that refers to a generic networking model (the OSI Reference Model) that postdates TCP/IP by several years. The layers and their names are shown in Figure 3.1.

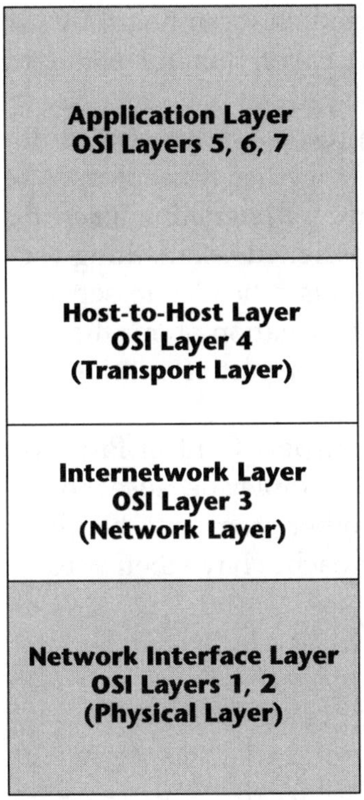

```
┌─────────────────────────┐
│                         │
│    Application Layer    │
│    OSI Layers 5, 6, 7   │
│                         │
├─────────────────────────┤
│    Host-to-Host Layer   │
│       OSI Layer 4       │
│     (Transport Layer)   │
├─────────────────────────┤
│    Internetwork Layer   │
│       OSI Layer 3       │
│     (Network Layer)     │
├─────────────────────────┤
│  Network Interface Layer│
│    OSI Layers 1, 2      │
│    (Physical Layer)     │
└─────────────────────────┘
```

Figure 3.1 The TCP/IP protocol stack.

Everything in the Application Layer is either specific to this host (such as a FreeCell game) or involves a logical relationship across the network between your host and another (such as your email program, where your host is a client for a server out there somewhere). The Network Interface Layer (often called the Physical Layer) governs how this host communicates with the network. Between the two is where TCP/IP really works—in the Internetwork Layer (usually called the Network Layer, or Layer 3 due to the popularity of the OSI Model's naming and numbering scheme) and in the Host-to-Host Layer (again, usually called the Transport Layer, or Layer 4).

We can review what happens to a chunk of information you need to send across the network from your host to another. You work with some sort of application, which creates a stream of information and sends it through the Application Layer, where a logical relationship between your host and the other is established. As with a telephone conversation or message exchange, you aren't directly connected to each other; you are communicating through intermediates, which are transparent to you. This is the same kind of logical relationship, or *session*, that is conducted between hosts. By the time this stream of data arrives at the Transport Layer, it is in a standard format and where it needs to go has been established.

The Transport Layer breaks the information into *segments* and multiplexes the information from several applications together. It keeps track of which information belongs to which application by designating a *port* for that application (in fact, most ports are already assigned—something very important to network security). Each segment gets a header to separate this segment from all the others and to carry information about the segment's content. Many people think of the header as a sort of envelope for the data.

Two protocols operate at this layer: TCP (Transport Control Protocol) and UDP (User Datagram Protocol). TCP is more complex because it operates with some quality control; UDP assumes another program is handling that and so does not waste effort on it. Both approaches have their uses.

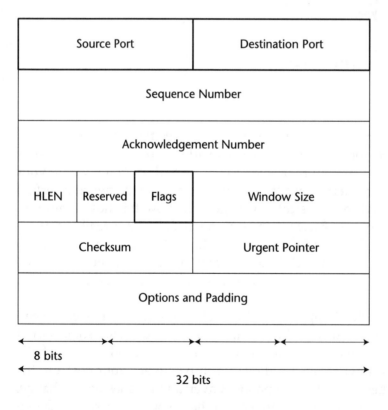

Min header size = 20 bytes (160 bits)

Figure 3.2 TCP header.

The TCP header, seen in Figure 3.2, is always shown in this format, due to some internal architectural issues that are not important here. I have emphasized the first two fields of the header, the Source and Destination ports, along with the Flags field. Certain ports have been designated by IANA, the Internet Assigned Numbers Authority, for certain protocols or functions. Numbers in these fields range from 0 to 65,535; the first block (0-1023) was assigned to the most fundamental protocols and functions with regard to networking. These ports are where many attacks are targeted.

To see which ports are open on your computer (we assume you have a Windows-based computer), open a command line or DOS window (click on *Start, Run* and type in "command" with no quotes, click OK) and enter the command "netstat -a" (again, no quotes). You will see all active connections, many of which (if you are on a network) may have a status of "listening." It is not important for you to know what each of these is at the moment; what is important to realize is that each open port represents a conversation your computer is having with the network. Each is also a potential entry into your computer.

The flags are a set of independent settings that signal a message type. TCP requires that there be a coordinated session between the two hosts that are communicating; no session, no data exchange. The session is coordinated by a three-part handshake. The first machine sends a SYN (synchronization) message; the second replies with an ACK/SYN (acknowledgment and some synchronization parameters). The first then replies with an ACK, agreeing to the parameters. Among the Flags fields is one to indicate a SYN, one for an ACK (both turned on makes an ACK/SYN), and a RST (reset) flag. The latter flag orders the recipient to reset the TCP session (abort it, and then it can try to reinitialize it). Attackers abuse these three flags, as we shall see.

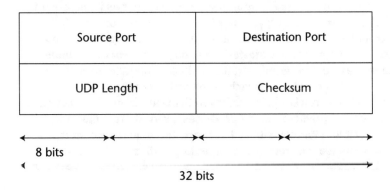

Max header size = 8 bytes (64 bits)

Figure 3.3 UDP header.

PROCESSES AND PRIVILEGES

Computers perform one action at a time, but their division of time is so tiny that it seems to humans that they are doing many things at the same time. Each thing is a separate *process*. Processes are not necessarily limited to activity on one computer. Multiple computers simultaneously working on the same problem via the same process are said to be sharing a *session*. To keep track of the session, a specific TCP or UDP port is used at each host. It does not have to be the same port on each host, though it often is.

As an example, some network workstations are designed (in hardware) to not know who they are until they have talked to a server—but how does the server know who's asking for its attention? A protocol called BOOTP handles this. The workstation is powered up and sends out an identification request from a BOOTP client, via UDP port 68; the BOOTP server replies with identification via UDP port 67. The TCP/IP protocol stack that comes up when the workstation is powered on knows these ports and what they mean.

These are examples of *well-known ports*. The well-known ports are the lowest-numbered ports, from 0 to 1023. As mentioned earlier, they are used for the most fundamental processes on a host and in host-to-host communications. They are the site of many network attacks because most of these fundamental processes require special privileges to access. If you can access the process, you can gain those special privileges, and those privileges may now allow you to manipulate other key processes in a cascade.

Different network operating systems use different names for these privileged accounts. The goal of an attacker against a UNIX system is to gain *root* privileges. Root is, essentially, the network god of a UNIX system. Other, lesser gods exist, such as *superusers* and *admins*, but root is superior to all. Root is allowed to change the most fundamental processes on a UNIX system. Root is normally a well-protected account because of the damage it can do to a UNIX system as a result of making a mistake in this arena. Among the processes root can modify is what activities the accounting function tracks—and that is one reason why root is a target of an attacker: It allows the attacker to erase any record that he or she was there. As a fundamental security measure, the root account should *never* be left named as root, much less use root as its password (for a login of root, password of root; you may hear this called "root-root").

In Windows NT and its follow-on, Windows 2000, the default godlike account is the Administrator. If the server is a master server on the network (a *domain server*), the account has Domain Administrator privileges. The Windows environment uses a more granular approach to the actions that a set of privileges grants, but if an attacker gains access to the Domain Admin account, he or she basically has all privileges on all processes, everywhere. When an attacker finds themselves in a Windows-based network, the Admin account is the first target, for the same reasons root is the target in a UNIX system.

The UDP header, seen in Figure 3.3, is much less complex because it does not perform all the quality-control processes that TCP does. But it also uses ports for the same purposes. To the maximum extent possible, IANA has assigned the same port numbers to UDP that it did to TCP.

When TCP or UDP has finished segmenting the stream of data and placing a header on each segment, it passes the segments to the Network Layer. Here, segments may be further broken into pieces, depending on how big a piece can be handled by the network en route to the destination (and that can vary for performance optimization reasons). The pieces are now called packets, and each packet receives another header, the IP header, as seen in Figure 3.4.

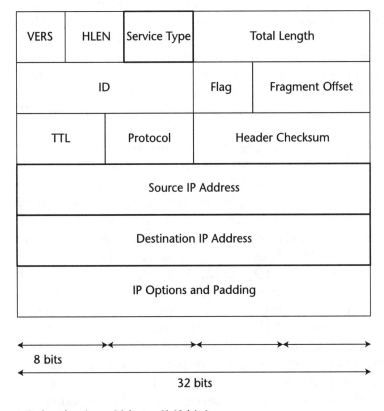

Min header size = 20 bytes (160 bits)

Figure 3.4 IP header.

The IP header has several fields that are not important to this discussion and three that are. The Service Type field is where Differential Services, or *DiffServ*, marking is applied. The Source and Destination Address fields are where the location on the network of your host and your traffic's destination is specified. Addresses used here are the familiar ones you see in the format A.B.C.D, like 192.168.1.105—and you can check your IP address by using *Start, Run* and typing in "command" with no quotes, clicking OK, then entering "ipconfig" with no quotes.

If you send a packet to another computer, the IP address you see here should be the address in the Source Address field (this can be altered, or spoofed). As a diagnostic check, you can find out if you can reach another host by sending it a ping (or pinging it—ping originally stood for packet internet groper); enter the command "ping hostname", using whatever the host's name on the network may be. A server called the Domain Name Server will translate the name into an IP address (if it can), and then it will test the connection to that host. "Request timed out" is a possible reply; this could mean a way could not be found to get there from here or back here from there—the reply is independent, or it could mean that the destination has been configured not to send a reply. You may also see a series of four replies because four requests are normally sent, and each will have an elapsed time in milliseconds. The replies name the destination's IP address.

Another message that originally began as a means to diagnose network problems is *traceroute* (the DOS and Windows command is *tracert*). This is a series of slightly modified pings, which returns the path taken by a packet from here to there. (Note: The path from there to here might be quite different; there is no requirement that the two be related other than at the endpoints.) Both ping and traceroute are a part of the Internet Control Message Protocol, or ICMP. They use a message called the ICMP ECHO_REQUEST and ICMP ECHO_REPLY. Because ping and traceroute are abused by hackers (see *Denial of Service/Distributed DoS*, later in this chapter), many sites have turned off any response to them, as noted previously.

Back to the Service Type, or DiffServ, field. While every packet has a source and a destination address, not all will have a DiffServ Code Point (DSCP) set. A total of 64 different codes may be placed in this field; 32 have been proposed as standard, 16 are reserved for future expansion of the standard (if needed), and 16 have been reserved for local or experimental use. These codes can be used to prioritize traffic on a network and to specify forwarding treatments. Their implementation on networks is relatively new, but they can be used to prevent abuse (from internal users as well as internal or external attackers) as well as ensure minimal transport times for special traffic (like Voice over IP).

The Network Layer has wrapped the packet in another header and now sends it down the stack to the Physical Layer, where it is encoded for transmission and placed onto the physical transport (copper wire or glass fiber). At the destination, the process is reversed, all the way up to the Application Layer. There is one more piece to understand about how your traffic gets there from here (and how the reply from there gets back to here).

It is virtually guaranteed that your packet will not get to its destination (and any reply from the destination will not get back to you) in only one step, even a giant step. At each step, or *hop*, along the way, the packet is passed from the Physical Layer to the Network Layer for a little change. This process is fundamental to how data travels through networks. Each interface on a network has an address, called a physical address (you can see yours with the command "ipconfig /all"). When traffic comes over the wire (or fiber), the interface examines the physical address; if it belongs to this host, even as a hop along the way to a different ultimate destination, the packet is accepted and passed on to the Network Layer, which places the physical address of the next hop on the packet and sends it back to the Physical Layer for placement on the wire. If not, the packet is discarded (like junk mail addressed to "Resident"). To get to the ultimate destination, your packet must be received and processed by many hops, each with a different physical address, each of which must be known.

A protocol called the Address Resolution Protocol, or ARP (which can also be abused by attackers) manages this. It builds a table, called the ARP cache, which contains mappings of logical interface addresses (IP addresses) with physical addresses. Based on that table, your packet is given the physical address of its next hop and sent along the way. The whole process is very much an automated assembly line for information: Pieces are added, and the bigger piece is sent to the next station, where it is processed. Meanwhile, this station does the same thing to the next packet, and the next.

Probes

In order for the attacker to get in and do something useful, he or she must have found the unlocked door or open window. To do that, the attacker has rattled doorknobs and tugged at sashes until he or she found one that gave. Just like a house burglar, why force a hard door when just down the way is an easy door no one locked?

Probes can be as simple as an automated ping. It is not hard to write a script, in any of a number of user-friendly languages. An automated ping script can be generated that simply modifies the IP address to interrogate by 1, then asks (essentially) if anyone is home at this address. After receiving a reply or a no response, the script adds 1 to create the next address and tries again.

In this way, an attacker can find the address of interfaces on your network that can be reached from the outside. That is why some interfaces have the reply to a ping turned off. (I have turned it off on my personal firewall.) An attacker may not want me to be able to trace back and find out who he or she is, so the attacker may keep my address in mind and keep exploring. Attackers often keep surprisingly good records, in order to use the same unsuspecting hosts as dummies or decoys over and over again.

Another form of probe is to use what is called a *war dialer*. This is a script, readily available, to dial a modem through a whole series of telephone numbers and to see on which ones a computer answers. One security consultant, to raise public awareness, ran a war dialer against the area codes in the Bay Area and Silicon Valley. He received modem replies—and could have logged into—computers from many of the Valley's most well-known technology companies. In fact, almost 18,000 modems answered, and almost 6,000 answered with enough information to determine the operating system (and version) on the host, which in turn tells an attacker which vulnerabilities are likely to be present.

A recent variation on war dialing is *war driving*, using a car to drive around while a wireless modem attached to a laptop scans for accessible wireless entry into networks. In an ongoing study, the results in the Bay Area for war driving are, if anything, even worse than they were for war dialing. Getting inside the network is half the battle.

When an attacker has found a way into your network and has several intermediate hosts he or she can use to reach you, the attacker may get more serious about probing and exploring your network. He or she will probably wish to disguise who and where he or she is, just in case you detect the attacker, so the attacker will employ *spoofing*. This substitutes a false source address in the IP header for the attacker's address. This basically involves lying to the application creating the traffic about what this host's IP address is; while not easy, spoofing is not terribly difficult, either. Another use of spoofing, by the way, is to arrange for the packet to arrive at your network with an address that belongs on your network. Of course, you allow your own family in through the door, don't you?

The problem with this is that addresses are relative to network location. An address from your network should be *inside* your network; how can a packet arrive at your ingress with an address that says it began inside?

That forms a loop, in which case network traffic circles through routers without ever arriving where it belongs. Routing protocols have procedures to prevent loops, so how did a packet that says it originated inside the network arrive from the outside? Good network management sets filters at your ingress to drop packets that purport to originate inside your network.

One probing tool the attacker will probably place in your network is a *sniffer*. In general, a sniffer is a software package designed to capture and decode packets, displaying their contents (which are normally not encrypted). It is used legitimately to troubleshoot network problems; however, a sniffer planted in your network by an attacker has typically been modified to copy certain packets passing by and store them in a designated location for later retrieval by the attacker. The material captured is often user identification and password. If your system sends these through the network unencrypted (because, after all, we're *inside* the network and therefore safe, you know), the attacker has hit the jackpot.

More likely, the passwords themselves have been encrypted while the usernames/logins remain clear text. This is not a problem; the attacker retrieves the stored file and can run any of several hacker tools against it. People have to remember passwords, so they choose easy ones, like "password" (the most commonly used password in the United States is, indeed, password). Another favorite is "lemmein" or "letmein" or a variant on this theme. Dictionaries are available to the hacker of the encrypted versions of these, of other popular passwords, and even of the entire English language, with variations of capitalization. One thing computers do better than people is tedious, mind-numbing work—like comparing one set of data to another. A few minutes to a few hours, and the encrypted passwords captured by the sniffer have been broken.

Now the attacker can log into your network and be accepted as one of your own. Depending on the privileges assigned to the account the attacker logged in under (a question resolved by some simple experimentation; the attacker will experiment until he or she locates the best account for his or her purposes), he or she can now go anywhere at will and do what he or she wants to your network.

Up to this point, while your network's privacy has been violated, no real damage has been done. Everything from here forward is damage.

Viruses

In biological terms, a virus could be considered an organism wannabe. It has portions, but not a complete set, of DNA, so it cannot replicate on its own. It needs a host organism to provide a friendly home. It enters a cell, captures control of the important set of existing processes, and orders the cell to do its bidding (make more copies of itself and disseminate them). This may have unpleasant, nasty, or even fatal effects on the poor host. In fact, most viruses are not fatal; more often they are annoying and unpleasant, or they may cause cosmetic problems (such as a wart). A virus that kills the host can only be considered "effective" if it does so after creating many, many, many replicas of itself and causing the host to disseminate them while it dies. This is not impossible, but it is a narrow path to walk.

A computer virus is a program wannabe. It is incomplete in its own right; when incorporated into another program, however, it is able to take control of that program and wreak computer havoc, should it be so written. Early viruses were mostly annoying, with screen takeovers that announced you had been had by *<insert boastful nickname here>*. Later viruses became more sinister, doing annoying things such as removing all files of a certain type from your hard drive or even causing serious damage by rewriting key portions of the hard drive so that the computer could not be rebooted without becoming completely unusable.

Virus writers have evolved, and so have the viruses themselves. The better, more effective viruses infect a host and remain essentially invisible. They operate in the background, performing the task set to them by their writer without degrading the host's performance enough to cause complaint. That task may be collecting information about users on the machine (logins and passwords), erasing any record of their presence (removing the audit trail of its operations that would normally be kept by the operating system), probing other addresses on the network for vulnerabilities, corrupting stored files, and so on.

Anything an ordinary program can do, a virus can do as well. The practical difference for you is that you intended the program to do something; you did not intend the result from the virus. In fact, you probably don't even know the virus is there until *something goes wrong*, quite possibly publicly.

Worms

Worms are viruses that have been made independent. They are fully-operational as is, and they do not need to capture control of an existing program to execute their commands. From an attack perspective, a worm may be better than a virus. Both viruses and worms can be used to bring down a network, sabotage or corrupt files, copy and store critical information for later retrieval, and more.

> **WHY CAPTURE KEYSTROKES?**
>
> Invoking a cryptography program often involves entering a key; alternatively, it might generate a key based on a passphrase (bigger than a one-word password, and bigger means mathematically harder to break). Without the key or the key-to-making-the-key, the encryption cannot be readily broken.
>
> Like virtually all things computer, the problem needs to be addressed in binary. A key that is 5 bits long has 2^5 (32) possibilities, but a 6-bit key has 2 times that, or 2^6 (64) possibilities. A 10-bit key has 2^{10} (1024) possibilities; a 128-bit key has 2^{128} (3.4×10^{38}) possibilities. This is simply too hard to do for anything but the highest national priorities, and then it requires, to be discreet, substantial effort.
>
> Of course, if you can capture and record for later retrieval the keystrokes used to create the key, you can recreate the key yourself and decrypt the targeted information easily. The crime-fighting establishment has reason to want this capability because Bad Guys don't typically play by Nice Guy rules. They are likely to encrypt any evidence that may be used against them in a court of law.
>
> Of course, because network attackers are Bad Guys, too, it may be reasonably supposed that they, too, have (or can readily obtain, through their own sources) this keystroke-capturing capability.
>
> Do you have an important announcement coming out, one that might affect the price of your stock? How much would it be worth to know the content of that announcement in advance? Do you have a hot new project, one whose critical information is all available in soft copy? Remember, theft of proprietary information is a common, and very expensive, form of attack by mature hands.

There does tend to be a subtle difference in use between worms and viruses. Viruses are used when the attack is focused on the operation of a particular system, such as email or encryption. As an example, on November 20, 2001, MSNBC and *The Wall Street Journal* reported that the FBI "is developing" (I take that to mean "has developed") software to insert a virus into a user's computer in order to capture the keystrokes used to trigger encryption of data and emails by that user. The program is currently known as "Magic Lantern."

With that in mind, worm usage has tended to be more general. Rather than taking down individual programs or hosts, worms tend to bring down entire networks (if the attacker's goal is disruption, a worm does the big jobs better for less). They are able to do so because of their autonomy: Once triggered, they require no help from any other program or process to do what they will. They spawn one or more processes, each of which spawns the same, and so on. The spawn rate may be slow, in which case there is a slow degradation in network performance as more and more power is used by the worm processes. If desired, the spawn rate can be faster, enough to create a rapid cascade of bottlenecks and breakdowns, leading to panic among network operators as their world comes crashing down around their ears.

The first (publicly known) worm was that unleashed by Robert Morris, Jr., in November 1988. It brought the forerunner of the Internet, ARPANET, to a crawl. It exploited two known vulnerabilities in the UNIX operating system, for which patches had been available. Due to buggy writing, it replicated much faster than Morris had intended. It brought down significant numbers of the computers interconnected via ARPANET: all told, about 6,000 computers were affected (approximately 10 percent of the Internet then). The Morris worm did this in approximately 36 hours.

Contrast that with the Code Red, Code Red II, and Nimda worms in the summer of 2001. On July 19, the Code Red worm infected more than 250,000 systems in 9 hours (of course, the number of connected hosts has grown exponentially since then; nonetheless, this represents an enormous cost, distributed across thousands of victims). This worm scans for open hosts, infects them, and the worms on the new hosts scan, send infections, and so on, in a cascade. Code Red II, which struck a few weeks later, was less extensive. Nimda, in mid-September, exploited vulnerabilities created by Code Red and Code Red II, used multiple propagation paths, and, in less than 1 day, affected more than 100,000 hosts. All three of these attacked vulnerabilities in Microsoft's Internet Information Server software, vulnerabilities for which Microsoft had made patches available months before the first outbreak.

From the first worm to the latest series, a similar situation presents itself: The attacker hits networks in known soft spots, spots for which a fix has been available, but has not been applied. According to CERT, as of CY2000, 99 percent of all intrusions resulted from the exploitation of known vulnerabilities or configuration errors where known countermeasures were available.

If the fixes are not yet applied on your network, what are you doing about it?

Trojan Horses

Trojan horses are usually just called Trojans. These are programs brought into a system via another (presumably innocent) program. They nestle themselves in, unnoticed, typically awaiting a preprogrammed time or a signal to be activated. The use of a signal is more common, as the attacker cannot count on a particular host having had its time set correctly (it should be set, but he or she doesn't count on such things). Besides, if the attacker is able to insert the Trojan, he or she should be able to get the activation signal in.

Trojans are often inserted as part of the probing process; among other things, they can be used to create back doors into your network. With a Trojan prewritten, the attacker can automatically scan and insert the script wherever he or she obtains access; the Trojan then runs silently in the background on the network, collecting information (logins, passwords, credit card numbers, etc.) and creating a storage location in which to keep it until called for. Perhaps the Trojan creates a new user account, preferably with the strongest privileges possible, then hides all accounting records of its activities. Depending on how long ago this happened, when you realize your network has been violated, the Trojan(s) may be a part of your backups, from which you would nominally restore your network to a safe condition.

Trojans are also used to plant logic bombs or time bombs. Time bombs are simple; they are destructive programs triggered at a certain time. Technically, a Trojan is supposed to be a malicious piece of software that you (unknowingly) place on your system, while a logic bomb was placed there by someone else (perhaps as part of a worm's payload). The distinction is hardly ever made by anyone except network security professionals.

Of course, the threat of Trojans and bombs can be used for extortion. If you seem unconvinced by the threat, a demonstration would probably be in order. In 1996, a disaffected network manager cost his (former) firm $12 million in damages and lost contracts due to Trojans he left behind that deleted critical files. (Guess who was responsible for the backups?) Back Orifice is a Trojan designed to completely take over Windows-based machines. Another Trojan disconnected users' dialup modem connections and reconnected them via a 900 number in Moldavia, leaving the users on the hook (so to speak) for the connection charges. And Trojans have been used to weaken encryption packages such as PGP.

Denial of Service/Distributed DoS

Trojans have become a tool used in another form of attack (along with worms)—the *Denial of Service/Distributed Denial of Service* attack. Denial of Service is self-descriptive. The attack does not do anything to your data per se, but it denies you and your customers access to it. A distributed Denial of Service attack is one that originates from multiple sources simultaneously. This carries a double impact: Not only are more attack packets coming in, but because they are coming from multiple sources, it is more complex to defend against.

DoS/DDoS (henceforth just DoS) attacks almost always originate outside the victim's network. Often, the launch point for any given attack or portion of the attack is itself a compromised (victim) host. Different flavors of DoS attacks have been tried, but they are really variations on one dish: a flood of packets intended to overwhelm. They may overwhelm the available bandwidth (clog the pipe), or they may overwhelm the processing power of the target (fill the manifold faster than it can distribute the flow among the outlets). Each is called a "flood" attack; they generally fall into one of three categories:

- TCP floods (SYN, ACK, RST floods)
- UDP floods
- Ping floods (ICMP ECHO_REQUEST and ECHO_REPLY floods)

These floods are a series of hundreds (at least; more likely thousands) of packets, all directed at their target, consuming bandwidth, especially on the last segment, where they all arrive at their target, and consuming the processing power of the target. Each packet, after all, is addressed to this host and must be processed, evaluated, and (usually) responded to. Even if the bandwidth is not consumed, the host may well slow to a crawl or crash completely due to the processing load inflicted on it.

The TCP/IP suite was designed in 1971 for a set of connected networks that was like the modern Internet only with respect to unboundedness. The few networks connected were military and academic; trust was inherent in the relatively small community, so there is no real difficulty manipulating packet contents, such as headers, to disguise the origin of a packet. TCP/IP is inherently not securable. Therefore, it is very difficult, if it turns out to be possible at all, to trace a flood attack back to the originator. The source address in the IP header has probably been spoofed if the packet, in fact, came from the attack originator.

On the other hand, if the attacker has used a Trojan, carried by a worm or a virus, to create an unwitting client from an unprotected host somewhere, the source IP address may well be valid. Unfortunately, it is the address of another victim, not your attacker.

Two more recent flood attack types are worth mentioning. One is an attack on name servers. When the IP address of a host is unknown, but the name is known, you can ask to be connected by name. TCP/IP sends a request to a name server to resolve the name into an IP address; when the reply arrives, your connection request is formatted with the correct IP address, and TCP/IP attempts to connect you. If name servers become disabled or overwhelmed, the Internet will be given orders of magnitude more difficult to use. For instance, you log into a server on your corporate network—do you know its IP address? Probably not; if you look at a Windows logon prompt, you request to log onto a domain, and in the window is a name, not an IP address. Without name resolution, virtually no one can find the other host they need. Your network will cease to network.

The second, more recent attack type is an attack using Internet Relay Chat, or IRC. Lest you think this is not an attack that could cause a problem for a business network, reconsider. First, any attack that affects the Internet may affect your business simply because traffic slows down due to congestion. If your customers can't get to your Web site, do they care that the problem is congestion on the XYZ network that their packets traverse to get to you (if they even know that; most won't)? Of course not. They will simply decide that it is too hard to reach you, and they will try your competitor's Web site. You'd better hope traffic to your competitor travels over the XYZ network, too.

Another reason why the abuse of IRC could affect your network is that some managers of geographically distributed organizations are now keeping a chat session going among their personnel ... all day. Whether this is sound management practice or whether it annoys the people who have to keep the chat window open, tying up resources on their computer and distracting them as they keep one eye peeled for something concerning them, is not relevant. Businesses are using chat to communicate; attackers are using the same chat to attack.

Sample Attack

Here is a simplified example of how an attacker might put the pieces of your network mistakes together to bring down at least a part of your network.

Using an automated scanning tool, an attacker probes your network, searching for an opening with which to enter and exploit a vulnerable host (see Figure 3.5). In fact, you have several vulnerable hosts, one or two on each LAN segment (why they are vulnerable is the subject of Chapter 4, "Murphy's Revenge").

52   Chapter 3

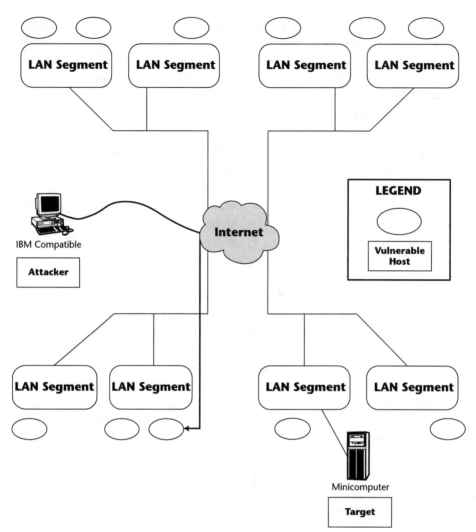

Figure 3.5   Network attack, Step 1.

The attacker gains access to a promising host and plants a Trojan, which implements a worm. The worm carries the Trojan as a payload, along with a few other things, as it propagates to every vulnerable host it can find (see Figure 3.6). It compromises them, and they, in turn, compromise every other host they can find. One of the "other things" the worm carries is another Trojan, an attack profile, which is now resident on every compromised host on your network.

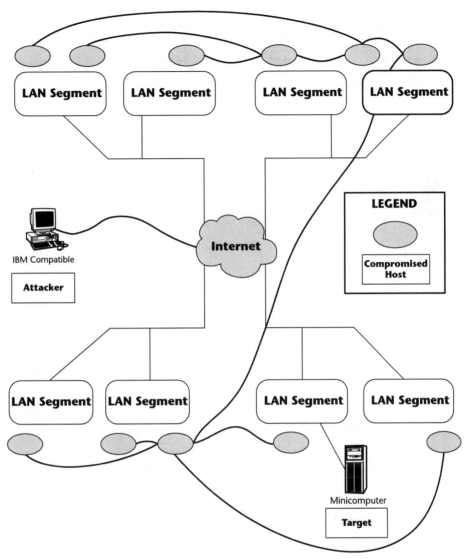

Figure 3.6 Network attack, Step 2.

When the attacker chooses, he or she sends the execute command to the compromised hosts, which implement the real payload (see Figure 3.7). They all send a service request, perhaps a TCP SYN message, to the targeted server. They don't send just one request each, however; they send that message hundreds, even thousands, of times each.

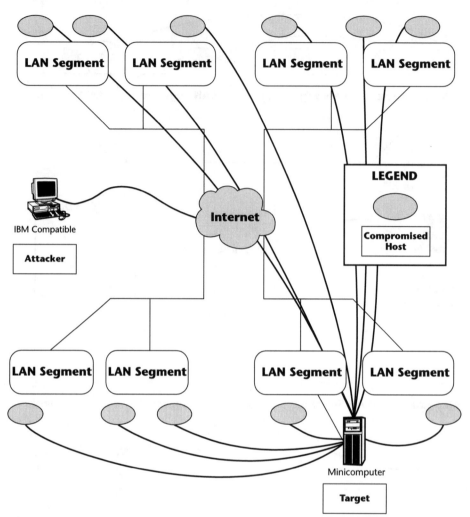

Figure 3.7 Network attack, Step 3.

A hurricane of traffic has just erupted on your network, and winds are battering the targeted server, which valiantly tries to answer each request. In the meantime, not only is the server not serving its normal clients, but enormous amounts of traffic are congesting your intranet, and no one is getting very much done.

Productivity is further affected by the time spent in multiple calls to your network administration team, who are less effective solving the problem than they could be if they weren't having to answer so many requests for information on what's wrong with the network and when do they expect it to be back up.

How difficult is it for an attacker to do this? Not difficult at all. An automated script exists for each step of the process, and those scripts are readily available via the hacker underworld. While this is a fairly sophisticated attack in its principles, the development of automated tools has meant that it takes no sophistication at all to perform such an attack. Detectives solve crimes, classically, by finding someone who had the motive, the means, and the opportunity. Motive was addressed in Chapter 2, "Network Threats." If we look at means and opportunity, the results are not encouraging.

Means

The means are readily available. Some automated attack tools known from experience in the network security world are the following:

- trinoo, in which an intruder compromises a few hosts (called "masters," via TCP port 27665), which in turn compromise many more hosts (called "daemons," via TCP port 27444), which flood UDP packets to randomized ports on the target(s) for a specified period of time.
- TFN, or Tribal Flood Network, which can operate with the same methodology as trinoo. In addition to UDP floods, TFN sends TCP SYN floods, ICMP ECHO_REQUEST floods, and ICMP Directed Broadcast ("smurf") attacks.
- Smurf attacks, which are ICMP ECHO_REQUEST packets with a forged source address. The packets are sent to the broadcast address for a network. (This is an address to which every host on the network must respond.) Because the source address is forged, the replies inundate another host rather than the one(s) that sent the request. This causes great network congestion and may crash the unfortunate recipient of all the ECHO_REPLY messages.
- Stacheldraht ("barbed wire" in German) combines features of trinoo and TFN, with the addition of encrypted communication between the masters and the agents (or daemons). It also updates the agents automatically.
- TFN2K, or Tribal Flood Network 2000, improves on the original TFN with more source address obscuration (to avoid ingress filtering), the addition of decoy packets to confuse troubleshooting, and the use of malformed packets intended to cause the receiving host to crash.
- mstream, which is similar to trinoo, but uses TCP ACK packets with randomized source and destination ports.

- t0rnkit (sometimes "tornkit"), which attacks UNIX hosts and first turns off logging, then installs several Trojans, including new versions of several critical system files, passwords those files with its own password, installs a sniffer and other tools, attempts to reenable any vulnerable processes that have been turned off, restarts all the new (compromised) processes, and then restarts logging. If the script is successful, the UNIX host is completely compromised at the root level.

I could go on, but there are a number of resources listed in the References, which will have more current information by the time this book is in printed form. While every one of the tools described here required a great deal of expertise and development to create, each one requires little skill to employ. Once a vulnerability has been discovered and a script developed to exploit it, the script may pass to the most immature of hands.

Opportunity

So much for means. Opportunity depends on how secure your network is. Security, in this context, is essentially an all-or-nothing affair—if there is one vulnerable host on your network, your network is vulnerable. The San Diego Supercomputer Center tested this concept by configuring a host with Red Hat Linux 5.2 with no security patches. The host was monitored, but otherwise not used in any fashion by the staff.

Within 8 hours of the installation, the first probe appeared; because the probe was testing for another operating system and vulnerability, this host was not yet compromised. By the time three weeks had elapsed, the host had been probed 20 times, for various known weaknesses of Red Hat 6.0; because this was a prior version it was not yet compromised.

On the fortieth day, the host was compromised. A rootkit (like t0rnkit) was installed by exploiting a vulnerability in the email system (the POP server), along with a sniffer, and (of course) the system logs were wiped. In less than six weeks, a new host was a back door into the network for an attacker. Had the host been running a more current software version, it would have taken far less than half that time. Why? Because it had not been properly patched for the known vulnerabilities of the operating system.

That is where we begin with Chapter 4.

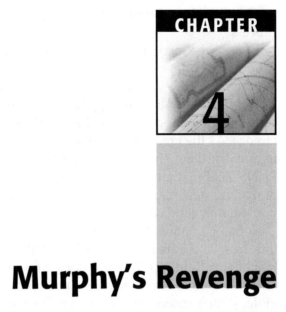

CHAPTER 4

Murphy's Revenge

Murphy's revenge: The more reliable you make a system, the longer it will take you to figure out what's wrong when it breaks.
Sean Donelan, on the NANOG mailing list

Sean Donelan's comment was part of a long thread discussing a massive failure of British Telecom's IP backbone on November 21, 2001. This was neither the first nor the last large network outage. Much of the discussion made reference to other "great" outages that have occurred; a continuing theme was the problems of adequately testing new software in the lab. No matter how many things you stress, the actual deployed environment is always worse. That's when the bugs show up.

System Is Not a Dirty Word

Or at least it needn't be. We grow up in the educational system, and we may think that is an unpromising beginning. Yet, at the end of our formal education, depending on the system in our childhood home, we really are rather well prepared for adulthood. Of course, we could be better prepared; we can easily see that, now—in 20/20 hindsight.

Some educational systems prepare children with a great store of knowledge, imparted with great rigor from a well-versed instructor. Others do less imparting and adopt a more free-wheeling approach, resulting in (perhaps) a less detailed knowledge set but a more flexible thinking process.

Systems are incomplete; they include trade-offs.

Perhaps a brief step back is in order—in Chapter 2, "Network Threats," I brought up the idea that a system has independent parts that operate interdependently to make a product or process that is more than just the collection of independent results. It is the interoperation that makes the difference. As the great science fiction author Robert A. Heinlein pointed out frequently, there ain't no such thing as a free lunch (often abbreviated as TANSTAAFL); that is the trade-off part. When NASA introduced the idea of making new planetary probes that were "better, faster, cheaper," wags immediately (and some would say, realistically) added "... pick any two."

Not all components can be maximized or even individually optimized; interference occurs (Carl von Clausewitz called this "friction" in the context of war fighting). When it does, you must choose which performance criteria are more important to you at this time, knowing that this set is fungible. Which criteria you choose will depend on which problems have a higher priority to your business—which is not necessarily the same as those you would choose solely for network performance optimization.

Before we can apply those principles to our network security, there is one more topic to address, reflected in the existence of systemic behavior in networks as well as individual computers. Bruce Schneier points out four important characteristics of systems: complexity, interaction, emergent properties, and bugs. We will meet these characteristics over and over again in the process of creating networks that can survive to perform their function. Individually, the properties are all partial reflections of systemic behavior; like the blind men and the elephant, each describes one aspect of a whole that is very much more.

Complexity

Systems are always complex. They have multiple parts, many or most of which have been characterized only by their independent behavior. The parts are put together in some sort of designed structure or structures, which, in turn, have intended input and output relationships. But because of a system's many parts, its final behavior is not always obvious; in fact, it is often not only obscure, but surprising (or even surprisingly obscure).

Interaction

Interaction is the norm when dealing with systems. Part of the reason for the unexpected behaviors cited previously is that the system's components interact in surprising ways. In a fractal-like behavior, the systems themselves also interact in surprising ways, creating a higher order of system with unexpected behaviors. This level of system then becomes a component in an even higher-order system (at some point, one is tempted to call it a supersystem, but next week that could be overcome by yet another overlay of a yet higher level of system).

As an example, we can move from a single computer to a LAN; the single computer as a LAN component has different characteristics (such as file sharing and the ability to cause another component to fail) from those it has as a component (single unit). The LAN becomes one segment of a building-wide network, which in turn is but one section of a campus network, which is a unit (connected by a WAN) of a continental corporate network, which is a portion of the global network.... We could also scale the connection to a service provider's network, which connects to the Internet. And that will someday be a component of the Interplanetary Internet (planning is already underway; see www.ipnsig.org).

Emergent Properties

Emergent properties does not mean that systems emerge full-blown from something else (like a moth from a cocoon). Rather, it means their behaviors emerge in unanticipated (which is to say, unplanned) directions. Inexpensive automobiles led to a change in the character of American cities—an emergent property that was certainly unexpected by early pioneers such as Henry Ford and Charles Olds. Competitive long distance telephone rates have led to a drastic drop in letter writing, which has significantly reduced the volume of first-class mail, necessitating an increase in postal rates to make up for the lost revenue. And, of course, air conditioning made Washington, D.C. tolerable enough year round to allow continuous governance at the federal level. Emergent properties are surprises; not all of them are pleasant. They result from the interactions of components operating normally but with unexpected interactive effects.

Bugs

Bugs, unfortunately, are the fourth property of systems. In a sense, a bug is an emergent property, but it is one that occurs from a design failure, rather than one that emerges when all components are operating as planned. To make problem solving more interesting, the appearance of the bug may depend on a complex set of conditions that cannot be exactly replicated. Without replication, it will most likely remain a mystery, casting its reporter in a dubious light.

In mathematics, this kind of behavior is often described as "sensitive dependence on initial conditions" and is a description of what has been popularized as chaos. Perfect predictability is possible if a complete description of the initial conditions is given. Unfortunately, that is generally either impossible or more expensive than the value gained from predictability.

Bugs then are inevitable. And known bugs (or defects, or deficiencies, or weaknesses) are targeted by your attackers.

Where Opportunity Knocks

Network administrators are often admonished by management to "fix the holes in the network." Unfortunately, there are thousands of possible holes to fix and limited time and other resources with which to fix them. The Common Vulnerabilities and Exposures (CVE) project at Mitre (http://cve.mitre.org/cve/) lists 1,604 accepted CVE and another 1,796 candidates for CVE status (in its update of September 18, 2001). The U.S. National Institute of Standards and Technology (NIST) maintains a metabase of CVE. As of its October 30, 2001, version, its searchable engine had records available on 3,095 CVE.

In the interest of putting the most effort where it will do the most good, the SANS Institute (www.sans.org), in cooperation with the FBI's National Infrastructure Protection Center (NIPC), developed a list of what it called the "Top Twenty" (originally the Top Ten, but more were needed)—the Twenty Most Critical Internet Security Vulnerabilities. These vulnerabilities are the target of a large majority of network attacks.

I prefer not to get in the middle of the UNIX-Windows religious wars; I find advantages and disadvantages to both networking systems. There are also significant security problems with both, as you shall see. SANS groups its Top Twenty vulnerabilities as General (of which there are 7 on the November 15, 2001, list), Windows (6), and UNIX (7). General vulnerabilities are not specific to a particular operating system, or OS. Another way to look at these is that they affect all systems. Even if your company has a pure Windows environment, you have at least 13 serious vulnerabilities to address. If you're pure UNIX, you have 14. And if, like most large businesses, you have some of each, you have 20 you must consider (at a minimum—remember, there are literally thousands more). An automated scanner is available (via hyperlink from SANS or at www.cisecurity.org) to check for these; once you have covered your existing systems, you may wish to set up a recheck schedule for any system that has access to the Internet (which is probably most, if not all, of your systems).

Top General Vulnerabilities

It makes no difference what operating system or combination of operating systems you run; you are vulnerable to these seven general vulnerabilities.

WARNING These are far more easily addressed when systems are installed for the first time than by patching afterward. If you choose to work at closing these openings after the fact, it will be expensive in system downtime. Of course, rebuilding a violated and crashed system is expensive, too. And the legal costs of sensitive information having been exposed when it should never have been accessible can be considerable as well.

#1: Default Installations

Default installations are designed to be easy and quick; easy makes it less likely that a user or an inexperienced administrator will leave out anything significant or that might be useful later. You could call it the kitchen-sink approach: Install everything they're ever likely to need or want, and throw in the kitchen sink, too, just in case. That way, when the customer needs another feature, it does not matter that he or she has lost (or misplaced, to be generous) the installation CD. With a custom installation, on the other hand, the user selects which features or feature packages to install. When others are needed, the user/admin must find the original CD and perform another custom installation. A standard or default installation is intended to protect these people from their own ignorance, but at a price.

The problem with this installation is that neither the user nor the administrator has any idea what is installed—and therefore what might need to be patched when a security update is announced. Default installations almost always include services that are extraneous to many users; those services often have open ports through which an attacker can enter.

Especially problematical are Web server installation scripts. Along with a large suite of services, sample scripts are often installed as well; these scripts are virtually never designed with the degree of care that the rest of the program receives. They are especially prone to buffer overflow attacks.

To protect against being vulnerable, even through a program designer's noble intentions, install only the minimum services required, and think carefully about what those really are. The Center for Internet Security (the same people who provide the joint survey with the FBI of the damage done by poor security) have compiled a consensus benchmark for a minimum security installation of Windows 2000 and Solaris. This is based on the real experiences of more than 170 organizations, from several countries around the world.

> **BUFFER OVERFLOWS MEET MURPHY'S LAW**
>
> A buffer overflow is self-descriptive: A buffer has received more information than it can handle. Buffer overflows are commonly used to attack systems and gain privileged access (this is a software design problem that should not exist, but does, in all operating systems).
>
> Murphy's Law, of course, tells us that whatever can go wrong, will. A corollary is that it will go wrong in the worst possible way. When an input buffer for a process overflows (for instance, it receive 536 bytes of data for a 512-byte storage space), it generally allows the excess data to overwrite the next space in memory. 24 bytes is probably not too dangerous, but the buffer overflow can be far larger than our example. In fact, *creating* buffer overflows takes careful programming based on detailed knowledge of the OS in order to get the right information to flow over into the right spots in memory. *Implementing* a buffer overflow requires running a readily available script.
>
> The result of a buffer overflow attack is often a completely compromised host—the attacker has gained the most powerful set of privileges available to this host. The highly skilled script creator rarely wastes time creating a tool that does not return the best possible value for his or her investment of time and knowledge.

If you are going to start revising your concept of how to install software, this is a good place to begin.

#2: Accounts with Weak/No Passwords

Yes, Virginia, such accounts still exist. The most common account password in North America is "password" (as previously mentioned). Many programs come with a default password on an administrator-level/root-level account. Hackers know these accounts and passwords, and those are the first accounts they test.

To check your vulnerability for this, you have to learn which accounts are available on your systems. Note that I did not say active, only available; when someone leaves the company, voluntarily or otherwise, is that person's account deactivated? Everywhere? Can you be sure, because you have good records of every account created on every host on your entire network? What about the default accounts created with the default installation script (which was run before you knew better)?

Look in places you might not immediately think of as a holder of an account: Anything that can be configured remotely can be accessed remotely and normally has a standard account for maintenance purposes. Check routers, printers, copiers, and so on. Printers are well known for providing superuser-level access to UNIX networks. You may wish to run a password-cracking program against your network: Several programs are readily available via reputable sources (SANS lists several in its discussion of this weakness). Use these with care, and with management's official support.

In fact, plan your password policy with management participation. I once interviewed with a company for a networking position; it was revisiting its password policy at that time. The company had had concerns regarding easy and old passwords and therefore implemented a 30-day life on passwords, made them more rigorous, and tolerated no exceptions. Unfortunately, the new policy was implemented for everyone on the same day; 30 days later the CEO ordered the policy scrapped when he couldn't get into his computer.

I accepted another company's offer.

#3: Nonexistent or Incomplete Backups

We all know we should make them. We especially remember, with a sick feeling in the pit of our stomach, when something goes wrong and we face the prospect of having lost some of our work. The only way to limit the damage from an attack is to be able to return to a status quo antebellum (literally, for those who can parse a bit of Latin to go with their binary).

It is not enough to merely make backups; you must know how to restore from them (when you desperately need the data is not a good time to be looking at the manual for the first time—if you can find it). Remember the properties of systems: They have emergent (unexpected) behaviors. You need to validate that you can indeed restore usable files, both program and data, from your backups before a crisis occurs. Otherwise, you have just increased the cost of that crisis by several factors, starting with recreating the data as best you collectively can after reloading all the software.

The backups should be off-site (that means, at the least, not in the same building or the building next door); it may be inconvenient, but it is safer if the data is at least 10 kilometers away, and this is a case where more is better. The backups, of course, contain the same valuable information as the originals: Protect them physically at least as well as the originals.

#4: Large Numbers of Open Ports

When I run the netstat command on a workstation on a large corporate network, I am not surprised to see a list of 12 to 15 open ports, usually listening. Identifying which ports are really necessary is more difficult. Ports 0 through 1023 are the well-known ports, reserved for privileged and other processes important to a network. Ports 1024 through 49151 are reserved by vendors for their processes. Ports 49152 through 65535 are dynamic and/or private ports.

Because hackers violate your network's privacy, it is not unexpected that they do not necessarily restrict their activities to dynamic ports. In fact, several attacks use signature ports in the registered ports range (for the latest on these ports, see www.iana.org/assignments/port-numbers). You should periodically scan the open ports on your network hosts. An interesting comparison occurs when you run netstat and then a port scanner against the same host; the results ought to be the same, but a port scanner will often find more ports open than netstat reports. Several port scanners are available via security Web sites; you must scan the entire range (0-65535) for both TCP and UDP.

Before scanning ports, again, have explicit permission (in writing is safest). Some implementations of the TCP/IP stack do not respond well to a scan. Likewise, if you have an intrusion detection system (IDS) on your network, it will notice and trigger an alarm. Depending on your firewall configuration, it, too, may signal an alarm.

Once you realize the number of open ports in your system, you will probably become very interested in closing as many of them as possible. That starts, just as the software installation thought did, with the idea of deciding what you need on this host and installing only that (or allowing only that port to be opened).

#5: Not Filtering for Correct Ingress/Egress Addresses

Ingress traffic is entering your network; egress traffic is leaving it. Why would traffic entering your network have an IP source address inside your network? It shouldn't—there is no reason for your internal traffic to loop outside the network and reenter. Likewise, there is no reason for traffic originating inside your network to have any IP source address other than one associated with your network.

If either of these events is occurring, someone is spoofing and using your network as part of the attack (you may be the honored target or just have one or more zombies—another term for daemons or slaves—on your network). Firewalls and access filters on edge routers (routers at the edge of your network, where you interface with someone else, like an ISP) should be configured to prevent passing such packets; they should also log the event and possibly trigger an alarm.

#6: Nonexistent or Incomplete Logging

Of course, I assumed you are logging events on your network. You should be. Logging requires not only storage (redundant storage in different locations; logs are a frequent hacker target), but logs must be reviewed. It does you little good to have a log that shows your network was first penetrated six months ago. What could you have done had you known sooner?

When you are attacked, your logs will help you trace and diagnose what the attacker did. Logs also help you detect the attack in the first place by providing information about the normal activity on the network (again, provided the logs are read). There are numerous sources on what should be logged and how often; to a large extent, it depends on what you do where on your network. And, because attackers will target your logs, replication across the network will help ensure their integrity. We will revisit this in the *Common Threads* section later in this chapter.

#7: Vulnerable CGI Programs

CGI stands for Common Gateway Interface. These programs provide the interactive experience people look for from a Web server; they are used in both Windows- and Apache-based systems. Sample CGI programs are often included with the server software; unfortunately, these programs can act as a back door into the server, as they are well known in their own right by the hacker community. Because CGI programs must be able to act on the server, they often carry extensive privileges, making them an attractive target.

CGI programs have been exploited to steal customer data (account information, credit card information, and so on). Having a secured transport system may ease customers' worries about their data, but, as we discussed earlier, if you can get into the server, it is much easier to locate the important data and copy/corrupt/erase it in situ. CGI vulnerabilities remove much of the difficulty from this process.

Remove any unneeded scripts, especially sample scripts. But then that should be part of not performing a default installation, shouldn't it?

Top Windows Vulnerabilities

The general vulnerabilities just discussed are related more to how components of the network are installed and managed than they are related to the quality of the component itself. The Windows vulnerabilities, like those in UNIX which will follow this section, are somewhat more technical. Rather than get bogged down in the details of these vulnerabilities, I will give them an overview; detailed descriptions may be easily found at the SANS, Mitre/CVE, and NIST metabase Web sites cited earlier.

#1: Unicode Vulnerability

Unicode is a standardized set of unique numbers for every character, regardless of the platform, program, or language. Most vendors, including Microsoft, have adopted the Unicode Standard. A URL (Universal Resource Locator, a Web-type address, such as http://www.yahoo.com) with a carefully constructed defect can force an IIS server to "walk up and out" of a directory and then execute arbitrary scripts. This is sometimes called a "directory traversal attack."

Microsoft has provided a patch for this flaw in at least six different updates, including Windows 2000 Service Pack 2. There is also an IIS Lockdown tool and a filter available to protect against this flaw. If you are current on your IIS patches, you should not be vulnerable.

#2: ISAPI Extension Buffer Overflows

Ah, another buffer overflow. The input buffer that is attacked in this case involves those supporting DLLs (Dynamic Linked Libraries) that support the Internet Services Application Programming Interface (ISAPI). This allows developers to create scripts extending the service capabilities of an IIS server. Like CGI scripts, these can be exploited. When the input buffer is overflowed, the result is often a complete compromise of the IIS server.

Again, Microsoft has made multiple patches available, including the Windows NT 4.0 Security Roll-Up Package.

#3: IIS RDS Exploit

Remote Data Service is a means of managing an IIS server remotely; it can be attacked and compromised. A version upgrade of the MDAC software should resolve the issue. Otherwise, while there are no patches per se, Microsoft does have instructions for correcting the vulnerability via a change in configuration; these instructions are available in three different security bulletins.

#4: NETBIOS—Unprotected Windows Networking Shares

Windows makes it relatively easy for a networked host to make directories ("folders") and their contents available to other users on the network. Users, however, often share their entire C: drive, which, of course, includes all the system software. And they make it available without even requiring a password. The actual file sharing is done via the Server Message Block (SMB) protocol; unfortunately, it can be manipulated to also make available critical user access information (username, last logon date (from which a hashed password might be located), Registry keys, and so on). Access is obtained via a "null session" connection to the NetBIOS Session Service.

The ports used by the NetBIOS null session can be blocked, anonymous connections can be denied, and rigorous file system permissions should be used instead of blanket sharing. Again, several security advisories describe how to protect against this vulnerability. Microsoft also makes available a tool that can scan a given host directly for exploitability.

#5: Information Leakage via Null Session Connections

Another name for a null session is an anonymous logon; it enables a user (local or remote) to retrieve information without being authenticated. Attackers capitalize on the fact that when Windows NT and Windows 2000 computers need information from each other (as a part of networking), they open a null session between them. And the system software asking for information has very high privileges with respect to the other host's system software. Attackers hijack this function to penetrate a host and gain privileged access to it.

The same null session ports can be blocked for external attackers (via your ingress routers and/or firewall), but you must not block them internally if your Windows hosts are to interoperate with their domain controller. A key setting in the Registry, however, may be modified to protect against this attack.

#6: LM Hash

LAN Manager is an outdated LAN software, but Windows NT and Windows 2000 have maintained a certain level of backward compatibility with it. One of the compatibilities is that it stores the LAN Manager hashed passwords on the NT or 2000 server. These passwords can be broken fairly easily because the LM hashing (conversion to an encryption output) is very weak.

Protection against this is a little more difficult, depending on your system. If you maintain Windows 9x (Windows 95, 98, or ME) workstations on your network, they limit how much you can remove the use of this older password format. If you have an all-NT/2000 network, you can restrict the password hashing to a more rigorous standard for use. The older, weaker versions are still created and stored in a known location unless you change a Registry key (this is available for Windows 2000 and XP only, but not NT). Several Microsoft Technet articles address this issue.

Top UNIX Vulnerabilities

Like Windows, UNIX-specific vulnerabilities are in the details of implementation.

#1: Buffer Overflows in RPC Services

RPCs are Remote Procedure Calls, the means by which UNIX computers cause a program on another computer to be executed. Multiple Distributed Denial of Service attacks have used RPCs to create their zombies. An attack on the U.S. Department of Defense systems exploited an RPC flaw.

Certain ports can be blocked at the firewall or ingress router, but you must leave them open internally. Three distinct RPC daemons are the most commonly exploited; you should see that they are updated with the most current patches. UNIX (Sun Microsystems, Hewlett-Packard, Compaq, IBM, and so on) and Linux (Red Hat, Debian) vendors have made them available.

#2: Sendmail Vulnerabilities

Sendmail is possibly the most exploited UNIX service. Partly, this is because it is everywhere there is UNIX—it runs the mail service—but mostly it is because sendmail has many and various flaws. Most of them have patches available, but attackers are still able to send a (carefully constructed) message to sendmail, which interprets it as a program to be executed; the program often requires sendmail to deliver a copy of the system's password file.

Sendmail must have the most current version possible and/or be fully patched. CERT maintains the latest information; search on the keyword "sendmail". While every host needs to run sendmail, it does not necessarily need to run the daemon mode version, unless the host is a relay or sendmail server. The daemon mode should be turned off unless actually needed (another nondefault software installation).

#3: BIND Weaknesses

BIND (Berkeley Internet Name Domain) is a DNS (Domain Name Service) package that is very widely deployed. BIND, like sendmail, is known to have multiple weaknesses. A name server is generally connected to multiple networks because it exists to resolve names to addresses for them. If the BIND service is compromised, the attacker is often able to gain strong privileges, which can then be used to install and operate automated scanning tools to probe the attached networks for their weaknesses.

Default installations of UNIX packages often include the BIND daemon; if you find the service is running, disable it unless this is actually a name server. If it is a name server, ensure that it is running the most recent version of BIND, including all available patches. Specific protection steps are available at SANS and from CERT; name service is critical to network operation, so this should be managed carefully.

#4: r Commands

A number of commands to execute programs in UNIX can be executed remotely (login and rlogin, for instance). r commands allow a single administrator (or a small number of administrators) to manage multiple hosts at different physical locations. As a convenience, trust relationships can be established so that the administrators don't have to log on with a username and password at each location—an attacker's dream.

Several r commands should be regularly scanned for; if an attacker gains access, one of the early steps they will take is to enable r commands (as well as delete any audit trail they know they are likely to have left).

#5: LPD

The Line Printer Daemon (LPD) is subject to buffer overflow. If it is handed too many print jobs in too little time, it will either crash or run arbitrary code with elevated privileges—which means the print server can be used to enter the network with strong privileges. Vendors have made patches for the daemon available. Likewise, disable the service if this device does not actually need to be a print server. LPD is often installed as part of a default installation.

#6: sadmind and mountd

sadmind is a System Admin Daemon that provides remote administration of Solaris hosts via a GUI. mountd is the file-mounting daemon. While both of these are technically RPCs, they are attacked so frequently that SANS felt it advisable to discuss them separately. They are frequently exploited because buffer overflows can result in attackers gaining access with root privileges.

sadmind and mountd should not be running on hosts accessible from the Internet. Vendors have also made patches available for these vulnerabilities. Further, file system configuration can be set to limit vulnerability.

#7: Default SNMP Strings

The Simple Network Management Protocol (SNMP) is aptly named. Its only means of authentication (remember, this is a protocol used to manage system devices) is an unencrypted community string whose default value is well known to be public. SNMP traffic reveals a great deal of what goes on in the network—that is what the protocol is for—but it reveals it to anyone who sees the packets, including those captured with a sniffer. Intelligence about a system only helps your attacker.

SNMP not only reports on the network, it can manage devices on the network. And it is unencrypted. And it is insecure. Is it any wonder that attackers can create SNMP packets to control network devices? SNMP is ubiquitous in networks, but it seems (thus far) to have been exploited mostly on UNIX networks.

If you must use SNMP, change the community string (which functions as a password) to something more like a real password. Filter SNMP at your ingress routers. Make your MIBs (Management Information Base packets) read only, if possible.

Common Threads

Some common threads emerge in these 20 most dangerous vulnerabilities. First, no matter whether you run Windows or UNIX as your OS, you face serious vulnerabilities in your OS. All systems are subject to buffer overflows, for instance. Buffer overflows, as noted earlier, are difficult to create but, once created and recorded as a script, are easy to use.

Another common thread to emerge is the danger of being lazy: lazy in your installations, lazy in your monitoring, lazy in your maintenance. OS manufacturers make default installations the path of least resistance; you must resist anyway. Default installations of any software are good long-term marketing—they deliver more than you asked for. Network security is a case where more is not only not better, it is very much worse. A phrase too often heard at CERT is, "But XXX isn't even on that machine!" Closer inspection reveals that, indeed, it is on the machine, usually as part of the default load. You purchased a software product for certain functions; install them and no others. If a function is not needed, don't put it there. This is a very negative attitude, but it is simply a reflection that prevention is uncomfortable for a little while, while the damage from no prevention can be permanently painful.

Lazy monitoring refers to not keeping logs, not keeping extensive logs, or keeping logs but not reviewing them. You will be attacked (you probably already have been, whether you noticed or not); good logs that are reviewed regularly will detect the attack (if it is subtle and has not manifested damage yet), they will help track where the intruder went and what he or she did, and they will help build a case for you in court, should you choose to prosecute. Logs are your eyewitness accounts; even if you don't use them for legal proceedings, they help you recover internally. Logs should be kept on all servers, routers, and relays. Logs should be written locally and *immediately* replicated to another host elsewhere on the network. This process should not be obvious; as an attacker almost always removes all log entries that he or she can find related to his or her activity. If possible, the replicated logs should be on a nonrewritable medium, so that, even if located, they cannot be corrupted from what was sent to them.

That brings us to lazy maintenance. There are two fundamental aspects to this problem, thought and deed. You must be aware of security issues in a timely fashion: Several patches exist for many of the major vulnerabilities discussed previously. Has anyone in your company thought about checking for security updates on your OS vendor's Web site (or are they too busy helping people reset forgotten/expired passwords?)? Does anyone in your IS department even spend time (as part of his or her job) staying current with or even getting better educated about security? The deed portion is that, having acquired and maintained knowledge, do you use it? Do you keep your systems patched with the latest updates? Have you applied all the patches to all the instances of that software everywhere in your network? It helps to maintain a database of what software was installed where; this need not be a burden—it is simply good accounting. Each site can maintain its local database and propagate it to a master (which is, of course, replicated as part of your backups).

If you have come to the conclusion that operating and maintaining a large corporate network is a nontrivial task, good. Too many corporate executives think IS is pure overhead and can always be trimmed when budgets need to be reallocated (especially in their favor). Your corporate network is your corporation's nervous system; without it, the brain cannot make the body deliver the business. With a compromised nervous system, the body responds to spurious signals.

If you can't afford to get the network right in the first place, when will you be able to afford to repair both the internal and the external damage?

Design Your Way Out of Trouble

It would be nice if you could design your way out of trouble, once you recognize you are vulnerable. Unfortunately, while design can both help and hurt, it is not sufficient in and of itself. Two major aspects of design contribute to your troubles or to their amelioration: topology and interface planning.

Topology

Network topologies are the structures of connections within your network. There is a physical topology (the actual wire/fiber/wireless/WAN connection structure) and a logical topology (which hosts are assigned into which groups). Physical topologies, shown in Figure 4.1, dictate how traffic actually enters, moves around, and exits your network. Logical topologies, however, dictate how traffic management *perceives* the network structure and then uses it.

Physical Topologies

The most basic physical topology is a flat one. Unfortunately, like the old strings of Christmas tree lights, wired in series, when one node goes out, connectivity to everything beyond it is gone. This makes it virtually useless for business purposes, though you may see an occasional flat section within a larger structure (systems within systems again).

A common physical topology is a star, or hub and spoke. In this topology, you gain control because all internode traffic must pass through a central node. You can thus filter and, if necessary, isolate a host from others (quarantine it). Unfortunately, the hub also represents a single point of failure; if it goes down, all internode connectivity is lost.

A mesh solves that by connecting every node to every other node with a point-to-point link. No one node's failure can cause a loss of connectivity to any other node; there is always an alternate path. From a business perspective, the reliability and availability of a mesh (especially a full mesh versus removing a few links and having a partial mesh) comes at too high a price: The links are underutilized and therefore too expensive for the value they bring. An alternative is a ring, where every node connects to both its nearest neighbors. If one node fails, it does not block access to others; redundant routes are available less expensively than a full mesh of physical connections.

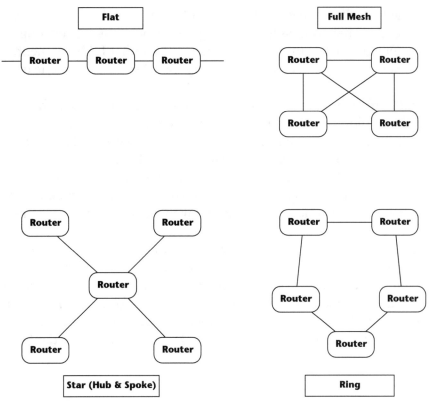

Figure 4.1 Physical topologies.

These examples showed connections between single nodes. In fact, physical connectivity in a network is often a combination of several physical topologies. A given LAN segment will typically be a physical star, centered on a switch. Several segments will be meshed (probably partially, enough to have redundant paths to the rest of the network). WAN connections are often part of a ring, which itself has dual paths (not over the same physical lines) between every node pair.

Logical Topologies

Figure 4.2 shows a logical tree structure. A true tree structure always descends from a single node, and there is a hub and spoke physical structure at each level in the downward direction. There is a strict hierarchical relationship between nodes at adjacent levels. IP addressing, properly done, follows this kind of scheme, with each level aggregating the addresses of the subordinate level. Traffic entering the network has a destination address. That address is sent along the branch that has the longest (most specific) match. For internal traffic to reach another LAN segment, it must traverse the logical structure to the very top (the root) and come back down.

In actual implementation, this is inefficient for large networks. Think of a global corporation. Traffic from the Atlanta segment may need to go all the way up to the corporate backbone to travel to Dar-es-Salaam, but it shouldn't need to do so to travel to Miami. Real networks look more like Figure 4.3, a blend of the tree structure with meshing for redundancy and a star topology at the base (or lowest levels). There is also redundant high-speed access to the outside world. This is far more common in large networks.

How does this help you secure your network from attackers? You will have routers defining each level in the overall hierarchy, and routers are configurable to help you protect against incursions and contain them if they do occur. Placement of the routers is a design choice, one not taken lightly lest your traffic bog down or loop around unnecessarily (like Atlanta to Baltimore to Reston to Raleigh to Miami versus Atlanta to Miami).

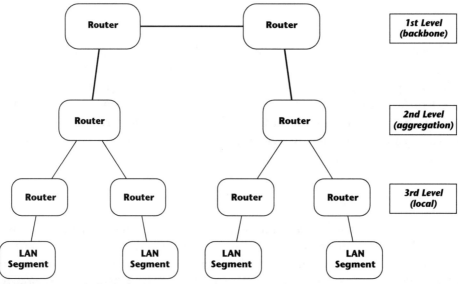

Figure 4.2 Logical topology.

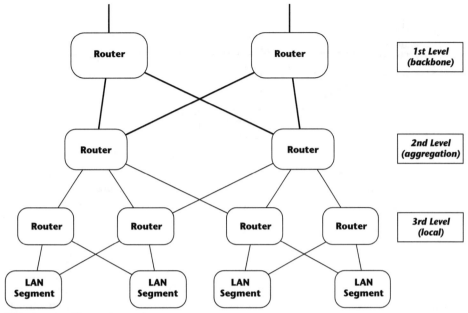

Figure 4.3 Network topology.

Defense in Depth

Routers and firewalls are your best defense. Most people think of firewalls as the first line of defense, but, at best, they are second (more likely third). Your first line of defense is a knowledgeable IS staff and employees who buy into your security policy, which we'll discuss shortly.

After your people, your routers are your next line of defense. Typically, a router faces the outside world. Behind the router, in an area known as a DMZ (Demilitarized Zone) are your publicly-accessible servers. Connecting the DMZ to your internal network is another layer of routers, and behind them is the firewall. This is shown in Figure 4.4.

Your ingress routers will direct public access traffic (traffic bound for the addresses of your publicly accessible servers) to the DMZ. Traffic bound for other addresses inside your network will be routed to the firewall. The publicly accessible servers need to keep logs and records, which must be passed inside to your internal network. Thus, the DMZ must be connected to your network, but those connections, too, must enter via the firewall.

76 Chapter 4

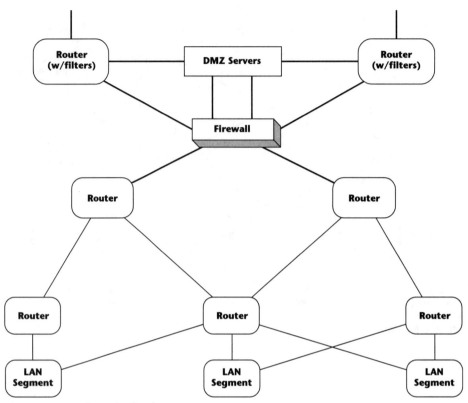

Figure 4.4 Defense in depth.

The external routers are your second line of defense (after your people, who are everywhere). Routers can be configured with access lists. These are filters against which every traffic packet is matched. Access lists have a simple principle: Anything not explicitly permitted is denied. Period. Router access lists can filter by a whole series of criteria; the most commonly used for this purpose are the source and destination addresses in the IP header and the source and destination ports in the TCP or UDP header. A fundamental filter here is to not permit traffic with your network as a source to enter from the outside; this is spoofed and you don't need it. Likewise, traffic with a source address of anything but your network is not permitted to egress, lest you be a part of a DDoS attack on someone else. This is a case where a gram of prevention is worth several kilos of cure; the legal liability of the network whose zombies attacked another has not been challenged yet (at least, not yet as this is written). You could be first.

Filtering by ports is a simple way to prevent attacks. For instance, TCP port 1025 is assigned to network blackjack by IANA. If network blackjack is going on, that's hardly an approved business use of your network. More likely, if traffic enters for that port, an attacker is using it, knowing that no one is likely to be looking at it for any reason. It is best to block all unneeded ports (that is, specifically permit only the needed ones); that is a level of detail, however, that many network administrators are not prepared to address.

It is not uncommon to group your servers logically on the network (all DNS servers on one segment, for instance—with redundant paths to that segment, of course). If so, you can apply a filter to routers on other segments to drop traffic using port 53 (TCP and UDP) and port 389 (LDAP, another directory service). Those segments with the ports blocked would now be protected against traffic abusing the DNS service (such as any UNIX hosts where BIND had not been removed).

Thus, router access lists provide defense in depth in another sense: They help protect you from abuse of services that were loaded on hosts by mistake (via previous default installations, for instance) or that are no longer needed and have not yet been removed. Table 4.1 lists the ports most often abused, arranged by service. This arrangement helps you block those ports on segments where such services are not used; however, if the service needs to be relayed through a router, the ports must be enabled on both interfaces traversed by the traffic (including backup paths to be taken).

Another item to be filtered by your access lists is the use of ICMP, the Internet Control Message Protocol. This is the protocol for network diagnostics such as ping and traceroute. These are abused (one attack is known as the "ping of death" as it is simply inundating a network with pings) and can also be used to diagnose your network's accessibility. ICMP messages travel directly over IP, and so they do not use ports. They do have message type codes, and you can filter on those. You should at least consider blocking incoming echo requests and outgoing echo replies (this combination means that someone inside your network can send a request out and receive a reply back, but someone outside the network cannot ping you and get a reply), and time exceeded and destination unreachable messages (you may wish to allow the "packet too large" form of destination unreachable—this message is used to correctly size packets for transit). SANS also recommends blocking any packets with source routing set or any other IP header options.

There is one further advantage of applying filters aggressively on your network, and that is your own internal traffic management. Your physical topology may make certain traffic flows possible that you do not find desirable from a performance standpoint. You can filter the undesirable traffic off those links. You can also manage traffic by configuring a router to handle traffic by DiffServ Code Point (DSCP); this allows you to ensure that certain traffic types receive priority treatment.

Table 4.1 Commonly Abused Ports

SERVICE	TCP PORTS	UDP PORTS	PROTOCOL/ COMMENT
Login or logon	21, 22, 23, 139, 512, 513, 514		telnet, SSH, FTP, NetBIOS, and rlogin
RPC and NFS	111, 2049, 4045	111, 2049, 4045	portmap/rpcbind, NFS, lockd
NetBIOS in Windows NT	135, 137, 138, 139	135, 137, 138, 139	for Windows 2000, add 445 TCP and 445 UDP
X Windows	6000-6255		GUI for UNIX systems
DNS and LDAP	53, 389	53, 389	relay routers should not be blocked
Mail services	25, 109, 110, 143		SNMP, POP, IMAP, but do not block relays
Web services	80, 443, 8000, 8080, 8888		HTTP, SSL, but enable to Web servers
Small services	0-20, 37	0-20, 37	housekeeping, time
Miscellaneous services	70, 119, 161, 162, 515, 179, 1080	69, 123, 514, 161, 162	TFTP, finger, NNTP, NTP, lpd, syslog, SNMP, BGP, SOCKS

The Price of Defense

The best things in life may be free (a moot point, but it makes good poetry); that isn't so in networking. There is a price to be paid for defense: either performance or the cost of performance. Your internal data transit time may be increased (performance suffers), or you must pay for more capable networking equipment (you pay more to get performance).

The reason is simple. To protect your network, you examine packets for more than just an address; you decapsulate every packet and compare it against a series of statements. In the process of sending information over a network, it is segmented at the Transport Layer (Layer 4), where it receives a TCP or UDP header and is passed down. At the Network Layer (Layer 3), it receives another header in front of the previous header; this is the IP header. At the next layer down, it receives yet another header and a trailer (which contains a quality-control check). At each layer, then, the previous layer's information is said to be encapsulated. In order to route traffic, it is examined by removing the last layer of header and trailer and reading the IP header that is now exposed.

With no filtering, the IP header is read and the router or other host looks at the destination address and (essentially) asks, "Is this for me?" If the answer is yes, the packet is passed on up the protocol stack for other layers to process, and the routing processor turns its attention to the next packet. If the answer is no (which is far more common), the router examines its routing table (technically, it looks at the forwarding table, a subset of the routing table that contains the best routes only, with no duplicates) to see if there is a known next hop to which it should send traffic for this destination. It writes the physical address of that next hop into the Layer 2 header that it creates as it passes the packet down for transmission, then turns its attention to the next packet. If the router does not have a path to that destination in its forwarding table, the packet is discarded, and attention turns to the next packet.

When you require the router to apply a filter, you increase the workload on the processor enormously. Traffic flows into and out of each of a router's interfaces. Each flow can be filtered according to a specific list. As an example, let's consider the ingress routers in Figure 4.4 (those at the top of the figure, connected to the outside world). Assuming you have a Web site for marketing and contact purposes on the DMZ servers, many of the ports and functions listed in Table 4.1 must be allowed. Every incoming packet must be checked against the filter, line by line. As soon as it matches a filter statement's condition, that statement's action is applied and the router turns to the next packet. If it does not match any statement in the entire list, the packet will be dropped. Because of this process, called an implicit deny rule, the filter must include permissions for the items you want to pass through.

Even the very best-constructed ingress filters are often 25 or 30 lines long. Every packet must be compared against the filter and a decision taken. Obviously, a well-constructed filter will place the most frequently applied rules for matching (both frequent permissions and frequent denials) as early in the list as possible, while maintaining the order that filters out unwanted traffic while permitting wanted traffic. This minimizes the load on the processor by allowing it to finish with more packets sooner. Expect to need routers with more powerful processors in order to handle the computational load you are imposing to protect your network.

Unfortunately, this is not a problem where you can extensively distribute the workload; you will want redundant ingress points, and each must be filtered. If your incoming traffic is balanced over the incoming links, each ingress router needs to be able to handle its regular load. For business reasons, however, you may well want each to be able to handle the full load. Potential customers do not consider it excusable that you had a router or a link go down; they see only that accessing your site took a long time (if it was possible at all) and they turn to your competitor's site instead.

Olive-Drab Networks

The ultimate in standardization is supposed to be the Army (in the United States, the uniforms were once olive drab cotton). You could simplify your protection requirements by standardizing your network, right? The answer is, "Yes, but ...." The costs of standardization must be weighed against the benefits; the net result will vary from one business to the next. And after we examine requirements for your network to survive natural and unnatural disasters, you may wish to reconsider how much standardization is desirable.

Benefits

Standardization allows you to simplify the overhead of your networking workload. If all of your routers are Model X from Vendor Y, much of the configuration can be set up as a script. As with the tools available to attackers, the hard intellectual work of a few can be implemented by the far more numerous, less-skilled people.

Likewise, identical workstations can easily be configured identically, probably from a script. Identical servers can be handled in the same manner. If you standardize on one OS, your security management problem is reduced by about one-third to one-half; you simply have far fewer vulnerabilities to track and fewer knowledge sets to maintain. To provide the same services, you may well need fewer software vendors (because you have only the single OS); that results in more simplification of process and knowledge.

If this is beginning to sound like the benefits of default installations, it does bear some resemblance. A major benefit of both is simplification, which means that fewer things ought to go wrong. But, just as you should be wary of laziness in host configuration, there is a similar warning against standardization throughout your network hardware and software.

Costs

The potato was the salvation of the Irish peasantry in the nineteenth century; it grew well in their soil and provided copious quantities of nutritious food to a people who had been lacking just that. From a small beginning, just six plants, the potato spread throughout Ireland and became the staple in the peasantry's diet.

Until the blight struck. Worldwide, the blight was not a serious problem; there were hundreds, if not thousands, of varieties of the potato plant, and only a few were vulnerable. Unfortunately, included in those few was the variety that had come to Ireland. In the space of a few months, famine developed. There is plenty of blame to go around among all the parties involved, including the British government, the plantation owners who continued to favor wheat for export, and the peasantry who did not turn to an alternate crop in time; no one, it seemed, grasped the magnitude of the problem until it was too late.

If you standardize, and if a network blight strikes your OS, all hosts are affected; none is immune. If you have a mix of systems, some will be affected, but others will not. In nature, diversity is the process by which species survive; some variations do better when conditions change. Without sufficient variety, whole species perish from the Earth. Your network probably won't perish, but larger portions may have to be rebuilt if you lack sufficient variety in times of environmental stress.

There is a further cost, and that is in the flexibility that increases productivity. Suppose you have a group that would be more productive using UNIX workstations and a particular CAD package; if you standardize on a Windows platform, their productivity will suffer. Alternatively, if you standardize on UNIX, the rest of the corporation may not be as productive because they will not necessarily have the software sets most productive for them. And there is interoperation with your partners and customers to consider, as well.

In all likelihood, you will have a mixed network, which means your personnel must stay on top of a broader variety of network issues. When you deploy new hardware or software (which you must do to keep the network capable of supporting the ever-increasing information transfer requirements of business), that is another set of knowledge for them to acquire—especially knowledge of the new vulnerabilities that come along. While that is a cost, it has the salutary effect of enabling your personnel to be aware of other possibilities than those "normally" done in a given OS.

Chapter 4

Converged Networks

Converged networks is a relatively new term in networking, and it still contains at least as much hype as content. The original (modern) communications network was the telephone network. It was engineered for certain performance characteristics, for a mix of technical, business, and political reasons (the last due to its position as a governmental or government-sanctioned monopoly). Data networks developed later and piggybacked on the telephone network's infrastructure. As data became a larger portion of traffic, separate data networks were developed. Their performance criteria were not quite so strict, partly due to differences in data and voice traffic characteristics.

When data piggybacked on voice circuits, the stricter voice tolerances had no bearing on data. The recipient at the other end of the line was another computer, of course. Converged networks currently refers to voice and data (again) traveling in the same network, but this time it is the network primarily engineered for data transport. For voice to travel over data circuits, some means must be found to meet the requirements of the human at the other end of the connection. To hold a conversation, the voice packets must arrive within latency and jitter parameters, and they absolutely must be in sequence.

It can be done. It is being done. But it is not necessarily easy.

Circuits or paths must be engineered to pass traffic with the least possible delay. One issue that has been overcome is head-of-the-line blocking. Even though many streams of traffic are statistically multiplexed onto one physical wire or fiber, only one payload may be loading through the physical interface at one time; a big piece will take longer to load and block the access of any other piece, no matter how urgent, until it finishes loading. Early Internet backbones (while it was still ARPANET) carried data at 64 Kbps. T-1s increased that to 1.544 Mbps; T-3s followed at 44.736 Mbps. Asynchronous Transfer Mode (ATM) was developed for the T-3 backbones with small cells (48 bytes of data plus 5 bytes of header each) as its packets in order to ensure that large data packets did not hog the interface and delay voice packets. The price of ATM was a large increase in overhead on the circuit because every cell had the requisite header, and a large data packet of 1500 bytes required 32 cells (32 5-byte headers plus 36 bytes of padding to fill out the last cell versus one 26-byte header and no padding; the math is easy).

DIFFERENT REQUIREMENTS, DIFFERENT RESULTS

Voice traffic and data traffic are fundamentally different in almost all defining characteristics. Voice traffic is very sensitive to delay, both the amount of delay (latency) and its variability (jitter). Total end-to-end delay on a voice transmission must not exceed 250 milliseconds (one-quarter second); in fact, users will complain about quality on a land line if delay exceeds 100-150 milliseconds (people expect less of cellular, so far). Jitter distorts pitch and must also be held to close tolerances to meet people's quality expectations. Voice packets are heard on the receiving end in real time; they must arrive in sequence to make sense. For reliability reasons, the voice network oversamples: it transmits more samples of digitized speech than are required to reliably reconstruct it at the destination. Voice traffic is therefore relatively insensitive to packet loss; it can afford to lose a few here and there.

Data transmission is, by and large, insensitive to delay and to arrival order. Packets will be reassembled at the destination to be passed up to a higher-layer protocol; the only impact of delay and arrival sequence is on the buffer size required to hold the information until all the necessary pieces for handoff are present. However, data traffic is very sensitive to packet loss. Consider the following two sentences:

> I cannot go home.
>
> I can go home.

Only one word is lost, but the meaning changes completely. Incomplete data is likely to give invalid information, so data transmission protocols require what is essentially an accounting function for the proper arrival of all pieces. Missing pieces are retransmitted; the rate of transmission is adjusted if too many packets are being lost. This reduces the likelihood of dropped packets needing retransmission. The data rate is dynamic, and so the time lag between packet arrivals may vary.

To accommodate requirements for minimal, predictable delay and sequential arrival of constituent pieces, voice networks developed using end-to-end circuits; once a call was connected, all the voice packets reliably travel along the exact same path, for the entire duration of the call (unless a link fails, of course, but that is quite rare).

Data networks, on the other hand, are entirely flexible about the path taken from here to there. During a given session between two hosts, packets may travel over different hops to arrive. One reason for this is load distribution; when redundant routes with the same operating characteristics exist to a single destination, many routers are able to distribute the packets between them, balancing the loads and making better use of the bandwidth. Because delay is not relevant for data packets, slightly different transit times are not important.

At speeds above OC-48 (2.4 Gbps), head-of-the-line blocking is no longer an issue, even with large data packets. With moderate-sized packets, it is no longer an issue at OC-12 speeds (622 Mbps). Modern network backbones are at least OC-48; OC-192 (10 Gbps) is widely deployed, and OC-768 (40 Gbps) was tested with live customer traffic in the summer of 2001.

With optical networking spreading closer to the LAN's edge, the separation of voice and data is no longer required for latency issues. Jitter could be a problem, but optical networking is already sensitive to that. The only remaining issue is sequencing—ensuring that telephony voice packets arrive in the correct order because they are real-time and no buffering and resequencing are possible (some voice packets, such as rebroadcasts and streaming one-way audio, can be buffered, and usually are). If sequencing can be solved, you can save money on your networks by returning to one network infrastructure again, rather than the two or more that have developed.

Two solutions for this are being implemented, the use of soft (they must be periodically refreshed) circuit reservations via the Resource Reservation Protocol (RSVP), and traffic engineering, the creation of circuits or paths through the network with different treatment; traffic is steered to the appropriate path. The treatment is specified by configuring each router's interface (automation is making this easier than it might sound) to recognize packets bearing a certain DSCP and forward them in preference to others. This is a Per Hop Behavior (PHB) system. You may recall that the DSCP is a setting in the IP header (see Chapter 3, "Tactics of Mistake," Figure 3.4). With DiffServ, there is still the issue of establishing a means to ensure that the telephony packets arrive sequentially; some sort of traffic engineering must be used.

The Catch

There is a further problem with converged networks: attacks. Telephony networks weathered a series of attacks from the phone crackers (known as "phrackers" or "phone phreaks"). Telcos responded with measures taken at the physical layer, in network design and protocols, to prevent what you now see happening with your data networks. Voice over IP (VoIP) is not, in itself, susceptible to attacks, but the IP network platforms on which it runs can be. One vendor's implementation, which runs on Windows NT servers, was severely affected by viruses such as Melissa. All VoIP networks can be congested with worms. Prioritization of telephony packets via DSCP is fine, until the DSCP is spoofed.

Some will offer IPSec as a solution, but be aware that DSCPs are outside the fields that IPSec encrypts, and this is on purpose. It may be necessary to modify the DSCP as a packet passes through a router in order to maintain the level of treatment received; if the DSCP were part of the encrypted field, the encryption check value would change, which would invalidate IPSec (this oversimplifies, but the principle is correct). Encryption will not protect DSCPs from spoofing and/or corruption by attackers. It also slows down the packet's transit through the router, due to the extra processing required.

Nor, to my knowledge, have attackers yet targeted the protocols used to manage the network (especially OSPF and BGP), though their processes and message structures suggest several means of attack.

Before implementing converged networks, a business needs to take a very careful look at the traffic requirements and how those requirements can be met under the stress of an attack. As an example, not directly related to network attacks, every state in the United States has a law mandating certain performance characteristics for telephone emergency 911 service. In Texas, the 911 call must be source identifiable to the street address and building; in other states, it must be identifiable to the floor or even the room/cubicle/numbered space. What happens when an employee moves his or her IP phone when he or she moves to an office in another building? In one vendor's implementation, a network administrator with the appropriate privileges must make a software configuration change in that phone number's profile for e911 service to report its location correctly (assuming the employee notifies IS of the move). In another vendor's implementation, the change mechanism is less straightforward (it is partially hardware driven versus entirely software driven).

What is your liability when the employee uses that phone to call an ambulance for someone having a heart attack and the ambulance arrives at the wrong location because of your network configuration? And, more directly related to attacks, what is your liability when a DoS- or DDoS-type attack targets the e911 system via your converged network?

That would not be an easy attack to program; neither is creating a buffer overflow. But if one can be created and scripted, why not the other?

Operator Error

When you can find no other explanation for why things went wrong, the explanation given is *operator error*, whether it be in the form of pilot error causing a plane crash or user error crashing a hard drive. The cause of the error can be sloppiness, inadequate documentation, or insufficient training. Any of these make the steps leading to the problem hard to replicate. That is one place good logging will help.

That does not mean logging individual keystrokes, except on especially sensitive items. However, logging can help you derive the cause of your network problem; knowing the cause is half the battle to prevent it from happening again. George Santayana had a point: Those who cannot remember the past are condemned to repeat it. He was not referring to networking, but it applies here as well.

Sean Donelan has a point, too. When you have created and maintained a high-quality system and something goes wrong, the cause will not be obvious. Repair will take time because there are no easy candidates for a cause. In fact, the cause may well be the convergence of several behaviors that, taken separately, are no problem. It is only when taken together that they become dangerous, but systems inherently interact, internally and (as components of a higher-level system) externally. When those interactions are the cause of the crash, they are devilishly difficult to track down.

In the meantime, business needs to go on. The measures you have taken to protect yourself from natural disasters will facilitate that a great deal. That is where we turn next.

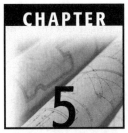

CHAPTER 5

"CQD ... MGY"

When anyone asks me how I can best describe my experience in nearly forty years at sea, I merely say, uneventful. Of course there have been winter gales, and storms and fog and the like. But in all my experience, I have never been in any accident... or any sort worth speaking about. I have seen but one vessel in distress in all my years at sea. I never saw a wreck and never have been wrecked nor was I ever in any predicament that threatened to end in disaster of any sort.

E. J. Smith, 1907, Captain, *RMS Titanic*

In April 1912, Captain Smith went down with his ship. It was intended to be his last voyage before retirement.

Of course, the sinking of the *Titanic* was not what many of us would consider a natural disaster. No one, presumably, would argue that it was a disaster: Approximately 1,522 lives were lost, an enormous number, especially at that time. Worth remembering, too, is that an immense capital asset, over four years in construction (two after her keel was laid down), was lost on her maiden voyage. And this disaster stemmed from human errors in the face of a natural phenomenon.

These next two chapters will focus on how businesses should deal with natural disasters: when your human actions collide with a natural phenomenon of deadly force. This chapter will lay some groundwork by examining the kinds of disasters people think of first, the large-scale natural events of tremendous force, and consider critically where you should be concerned with them and how likely they are to occur. Chapter 6, "The

Best-Laid Plans," will spend considerable time on how to be prepared for natural events that may affect your network's operations. The unspoken assumption at this point is that you will have some warning and can act in a planned (and presumably coherent) manner. The procedures developed for situations with warning will be bent when dealing with no warning; properly constructed, however, they will not break. We will focus on that in later chapters.

As we look at human error and natural disasters, a surprising number of lessons can be drawn from the calamitous maiden voyage of the *RMS Titanic*, and there may be a few myths to dispel as well. We can start with some facts about the ship, her planning and construction, and what happened when the business of a luxury liner collided with an elemental natural object.

A Classic Disaster

The *Titanic* and her sister ships, the *Olympic* and the *Britannic* (whose name had been planned to be the *Gigantic*, a plan changed after the *Titanic* sank), were intended from their inception to be the largest objects afloat. This was, of course, in a period when Great Britain and Imperial Germany under Kaiser Wilhelm II were engaged in a deadly naval arms race, focused on ship size as much as numbers. The three ships were not only intended to be the biggest, but also the most luxurious—the ultimate, as it were, in transoceanic transportation.

All three ships were designed to handle a great deal of damage, though they were never advertised as "unsinkable" (*Shipbuilder* magazine did call them "virtually unsinkable"). A frontal or a quartering ship-to-ship collision, the most likely serious hazard, might open two compartments to the sea; they were built to withstand three of the five forward compartments being flooded, or even all of the first four. No one ever imagined a sliding collision that opened five. When she flooded in the first five, her bow settled, allowing water to spill into other compartments aft of the first five (the water-tight bulkheads may have been left open to facilitate pumping). In a similar fashion, *Britannic* went down by the bow in less than an hour after striking a German mine during World War I; *Titanic* actually lasted around two and a half hours.

There has been a myth that no thought was given to having sufficient lifeboats aboard. In fact, she carried 20 lifeboats, 125 percent of the British Board of Trade regulations requirement for any ship over 10,000 tons displacement. *Titanic* displaced over 46,000 tons; regulations had not kept up. Two of her collapsible lifeboats actually washed off the deck when she sank; they had not even been deployed. Most of the early lifeboats were only partially filled, some less than half. The bridge officers and crewmen were apparently unaware that the davits had been tested with more men (70) aboard than the rated capacity (65); women and children, of course, weighed even less. The deployment was clumsy, reflecting the lack of drills for such an event, but then, no one could imagine that the lifeboats would really be needed. The North Atlantic was heavily traveled, and ships would be nearby to effect a rescue.

A ship, the *Californian*, was indeed within 19 nautical miles (or 21 statute miles/33.8 kilometers) of the *Titanic* and could see her lights from her superstructure (bear in mind that the horizon is 12 miles/19.3 kilometers away at the sea-level surface). Her captain had prudently halted in the face of the numerous warnings about ice fields, but he could afford to—they were traveling from London to Boston with no passengers. *Californian* did send a voice message to the *Titanic* less than an hour before the collision, reporting that they had stopped and why; *Titanic*'s radio operator brushed them off because he was too busy sending backed-up passenger telegrams (the set had been malfunctioning earlier). This was one of at least six ice warnings *Titanic* is known to have received that night. The bridge watch from the *Californian* could not tell in the middle of the night what type or size of ship they saw (nor, therefore, exactly how far away or who she was). The wireless operator had finally gone off watch and shut down the set; a bridge officer who liked to fiddle with it did not know how to power it up (wireless was quite new; most information was still sent in Morse code, as opposed to actual voice transmissions). The *Californian*'s watch later observed a series of white rockets from the mystery ship, but no colors—and colors had meaning: they identified the line a ship belonged to. When Captain Smith realized the gravity of the situation, he ordered a distress signal to be sent on the ship's wireless; the signal "CQD" was sent several times, followed by "MGY." "CQD" was the Marconi company's attention call for distress. Marconi was the actual employer of most British shipping radio operators; "SOS" had only recently been adopted as an international standard distress call at the recommendation of the Germans, for its easy recognizability, but the British shipping industry was loath to adopt anything German ("SOS" was actually sent in the final distress calls, just before the bow went under more than two hours later). "MGY" was the *Titanic*'s call sign. The distress call also included her position, but they miscalculated and reported a location over 13 miles from where she actually sank. Until daylight came, survivors could not be seen. The rockets were fired from the deck by a Quartermaster, apparently in an attempt to get the attention of a ship *Titanic* could see to the north (the *Californian*). When *Californian* finally roused its radio operator—well over six hours after the accident—he was able to learn quickly what had happened. Now in daylight, the *Californian* picked her way through ice for an hour to arrive at *Titanic*'s reported location, where they found another ship also searching; she then worked her way over to where *Carpathia* was just pulling in the last of the lifeboats after running flat-out for four hours in response to *Titanic*'s mayday.

Finally, humans behaved like humans in this disaster. Despite the orders to evacuate, the early lifeboats left well below capacity (several at less than half capacity) because people didn't want to leave the apparent safety of such a big ship for such a little boat. Not all men graciously sent women and children first. The lifeboats rowed vigorously away from the sinking ship, lest they be pulled under by suction (*Olympic* had been found at fault for a collision seven months before in which the water movement caused by suction from her 46,000-ton hull's displacement forcibly drew a 7,350-ton Royal Navy cruiser into her in a hard collision). Once the stern of the *Titanic* had joined the bow beneath the ocean, only one lifeboat returned to pick up swimmers (the others said they feared they would be swamped, even though they were capable of carrying twice as many people and they knew the swimmers would die).

Lessons from Failure

I have never paid much attention to the story of the *Titanic*, beyond seeing a few movie clips (which usually included the orchestra heroically playing *Nearer, My God, To Thee* as she went down). But Captain Smith's statement embodies the comfortable sense we all tend to have that no major disaster is going to happen to us. Afterward, of course, we all know better. The loss of the *Titanic* resulted from multiple failures, failures that occur in many catastrophes, natural and man-made. Here are a few I believe to be important.

A Trophy Property

The *Olympic*-class ships (named after the first one launched) were intended to be the elite of the sea lanes. They were to be symbolic of British dominance of shipping, the best in every respect. As a result, their design suffered somewhat from form over substance. For instance, the original drawings showed 48 lifeboats (which would have carried approximately 3,100 people), but White Star Lines rejected those drawings because her boat deck looked too cluttered. And, as it was, the 20 she eventually carried (capable of carrying 1,300 of the approximately 2,200 on board) were significantly more than British regulations required. The fact that the regulations were woefully out of date and that architecture had advanced to a size four times that of the largest conceived of in the regulations, does not seem to have been considered.

Many businesses invest in trophy properties, with distinctive architectures to be instantly recognizable. The Sears Tower in Chicago (with the double antenna towers rising into the sky) is unique. The World Trade Center in Manhattan was instantly recognizable, a symbol of New York, capitalism, and the United States. The Petronas Towers in Kuala Lumpur have a beautiful and distinctive silhouette. On a less dramatic scale, we have arenas and sports stadiums, corporate headquarters and local landmarks, all designed primarily to be identifiable. How far down the requirements list was safety—and safety in the face of what?

The *Titanic*'s plans overengineered the requirements for the worst disaster her designers imagined, a collision with another ship. That was the expectable sort of physical stress her construction had to be able to withstand. They apparently did not consider the fact that she was intended for the North Atlantic run between Europe and the United States, a route known for severe winters and spring ice. Possibly they did not extrapolate from the presence of ice the likelihood of a collision with ice. The *Titanic*'s entire sea trials consisted of one-half day, during which the engineers did discover that it took her more than 3,000 feet/914 meters to come to a stop when the engines were reversed from a normal cruise. Steering response and the best way to obtain the maximum directional change (such as reversing one inboard screw and running forward with the other in conjunction with rudder movement) do not appear to have been trialed.

Buildings do not need trials under motion, at least not of that sort. But what sort of modeling did go into the design of your trophy headquarters? It may well stand up to a severe storm of the normal variety; what about a hundred-year storm? What about natural disasters that have historically occurred in this area, but have not occurred recently? As an example, in 1811 and 1812 a series of massive earthquakes struck near New Madrid, Missouri, along the Mississippi River. The native tribes spoke of a day when the river ran backward. One temblor was so violent that church bells rang in Boston—almost 1,100 miles/1,750 kilometers away on a direct line (with low mountains between). These earthquakes are estimated to be among the most violent, if not the most violent, in North American history. The fault system has been relatively quiet ever since.

Much construction in cities along the Mississippi has been along the riverfront, for its scenic and historic value. The land underneath is a maze of old riverbeds; while that has been accounted for, experimentation after one of Mexico City's more disastrous quakes has demonstrated the remarkably fluid behavior of such material during an earthquake. The river valley cities have no building code equivalent to that of California, or Mexico City, or Istanbul, or Tokyo/Osaka/Kobe.

We recognize after the fact that we should have considered the unlikely event in our designs.

Warning Noted...

... now leave us alone. That has been heard more than once and emphasized afterward with bitter condemnation for the lives lost when people could have been evacuated in an orderly fashion, with time to spare. Aside from the public relations nightmare that would be for your firm, how would you replace those people and the corporate knowledge inside their heads? How would you sleep with the unnecessary deaths of those in your charge on your conscience? The excuse given is that most of those warnings are false alarms; what the excusers mean is that it's too embarrassing to have responded to a false alarm. Your people would probably prefer to be alive, and they would forgive the embarrassment.

Humans, by and large, assume the glass is half full rather than half empty: They interpret things in the most favorable light. The *Titanic* received at least eight warnings that we know of from other ships, one each on the two preceding days and six on the day she sank, warnings that culminated in one that described a rectangle of the sea with "... much heavy pack ice and great number of large icebergs, also field ice"—a rectangle she was already inside. The last six warnings came at 0900, 1140, 1342, 1345, 1930, and 2140. She struck the massive berg at 2341, two hours after the last warning—and still inside that rectangle.

Many natural disasters strike with little warning, but for many we do have warning. If a hurricane approached a coastal city where you have a major center, do you know that your personnel in charge would heed the warnings to evacuate, or do you recognize that they would delay as long as possible, on the chance the hurricane would veer away (as they sometimes do)?

Train the Way You Will Fight

That was a motto when I was in the U.S. Air Force. It meant to practice the way you will need to perform under pressure because that is how you will actually perform; you will behave as you have practiced. Likewise, when I was involved with the Olympic Development Program in youth soccer, the desire was to find players who could perform "at speed, under pressure from an opponent, in a game." Anybody can look good in practice. When you're going full speed, you're under pressure from someone who isn't your friend, and it certainly isn't practice; performing up to par is much harder to do.

To be prepared for disaster, you must plan what you will do, and you must practice the plans. Lifeboats were deployed clumsily because the *Titanic*'s crew had not practiced how to work the davits. They were hesitant to fully load the lifeboats because they thought the davits might buckle under the weight (in the course of practice, someone might have raised the issue and learned the system had been tested in an overloaded condition). In fact, the disaster occurred on a Sunday; after officiating at religious services at 10:30 A.M., Captain Smith declined for reasons we do not know to hold the usual Sunday lifeboat drill for passengers and crew (this was her first Sunday in service). Even one practice could have made a difference, especially if it had been that day.

Some will object that no one takes emergency drills seriously. If the CEO does, I think you will find that attitude suddenly contagious, at least on the surface, and performance when it matters will reflect that there had been some practice, however lightly taken by a few. Leadership works best by example; if you don't have time to muck about with safety drills, why should anyone else? Do as I say and not as I do?

What Did You Say?

Communications go south in a crisis, at an unbelievable speed (which also makes practice important). The important message, the most relevant content, may not be understood by the recipient because he or she has no idea (yet) that anything has gone wrong. On September 2, 1981, I was in HQ USAFE, Ramstein AB, Germany, when the Red Army Faction bombed the building. I took care of a few things that had to be done (training kicking in), then ran out of the building and across the street to the Officers' Club, where I telephoned my husband at his office at Sembach AB nearby. When he answered, I rushed out the words, "I'm okay." He paused for a moment, then replied with a long, drawn out, "Oka-a-a-y" of his own because he had no idea at that point that anything had gone wrong. I explained very fast that HQ had been bombed; he acknowledged that and dropped my call to alert his base (as he should have—terrorism was a real threat we lived with).

Preplanned messages have content that is expected (even if not right now), which makes them more readily understood at a moment when time can be critical. The signal rockets from the *Titanic* were a general get-your-attention message to a mystery ship to the north; the *Californian* saw them, but her captain asked repeatedly if they had any color because different colors represented different companies (and the only White Star Lines ship anywhere near would have had to have been the *Titanic*). Had they known the type and size of the mystery ship to the south, they could have calculated the distance, they could have roused the wireless operator sooner and perhaps saved many lives from the below-freezing water. After all, rescue by other nearby ships was the main reason no ship needed many lifeboats.

And, of course, the message content must be accurate. With sea-level horizon at 12 miles, a misreported position of over 13 miles meant that the first ships to arrive, when it was still dark, arrived at the wrong place to do any possible good. It was not until daylight that the correct location could be seen, by lookouts in a crow's nest well above the surface.

A Scarcity of Heroes

That is not a collective noun (like a gaggle of geese or a caucus of crows), though perhaps it should be. Heroes are scarce, and most people, while they might vaguely wish to do The Right Thing, are not at all sure what exactly it is, much less how to do it.

Don't count on heroes to save your people or your business.

And expect your people to act like people in a crisis. Without planning and practice, they won't know what they should do, so they will, in general, revert to what they do know. And, in the end, most will try to save themselves. One man in a lifeboat from the *Titanic* claimed he put his wife and children aboard another boat first; they were found after the fact aboard a boat loaded 15 minutes after his—and she divorced him.

Lessons from Success

There are always lessons to be learned from failures, but there are also lessons to be learned from successes. To that end, it is constructive to look for those who have handled disaster well, then look at what they did and why; their what may not always migrate, but if we understand their why, we may be able to apply some of their techniques, at least in a modified form.

One aspect to remember as we deal with natural disasters is their scale and scope; the extent of the damage experienced is often extraordinary (literally: out of the bounds of the ordinary). It is therefore beyond the capability of the ordinary authorities to handle. This may be due to any of the following or a combination of these: damage to infrastructure, preventing the ordinary flow of goods and services (how much food is in your house right now?); displacement of people, with the attendant problems of food, water, and shelter; and the displacement of businesses, which might otherwise be able to help with the displaced people.

When some sort of natural disaster occurs, we look to certain people to get a handle on the chaos, to restore the orderly functioning of society and enable business and people to recover and rebuild. Those most often turned to for this (rather than for charitable works) are the military. In many countries, it is the Army (or, along the coastline, the Navy); in the United States, it is often the National Guard, the Army organization of local citizens, charged primarily with local work, but available for national defense in an emergency.

They deploy, take over, and generally (though not perfectly) restore order and organization in an area where the ordinary means of government lacks the tools, the skills, and often the sheer manpower. An important distinction should be made here: Throwing bodies at the problem is not all the military does in this situation. If that were the case, every time a Third World army moved in and occupied a city, for instance, there would be perfect order and no reason for refugees to flee. So what does the Army or the Guard provide that a mob of men acting more or less in concert does not? Among other things, organization, training, attitude, and a plan. Do they always use those things perfectly? Of course not, but they get most of them right, most of the time. And believe it or not, their leaders spend most of their free time thinking about ways to fix the rest of it.

Organization

The Army (and, by extension, the local militia or Guard units) is supposed to be the epitome of over-organization, smothering the creativity of individuals who are honestly trying to get something done, albeit unconventionally. It often operates in just that fashion; the standardization that makes it so effective at the macro level may well stifle creativity at the micro level (though every army has organizations where creative individuals can thrive, once they prove themselves).

Nonetheless, a military organization has a defined and well-known hierarchical structure: the 603 TCS (squadron) belongs to the 601 TCG (group), which belongs to the 601 TCW (wing), which is a part of 17th AF (numbered Air Force), which is operationally a part of USAFE (United States Air Forces in Europe). Likewise, an army platoon is a part of a company, which is a part of a battalion, which is a part of a regiment, which is a part of a division, which is a part of a corps, which is a part of an army operating in a theater. At every level, relationships and authorities are known at all times and under all conditions; fallbacks are prearranged. The structure forms a system.

If conditions change and new ones are in place, relationships may change; a division may be reassigned operationally from one theater to another, but it knows there is a chain of command in the new theater and what it is, who can say jump and have to be obeyed.

More than anything else, the military's organization is about always having a defined structure within which people can get the job done, whether it's patrolling a Demilitarized Zone for infiltrators or reestablishing communications for a city after a Force 5 tornado has destroyed the civilian system. You don't need the discipline of boot camp in your company (no military organization maintains that except in boot camp, a specialized operation to change fundamental behaviors radically in a ridiculously short period of time), but the structure and relationships and fall-back system of authorities should most definitely be a part of your plan to keep your networks in the survivor column instead of among the list of casualties.

Training

A tired soldier is a happy soldier, so the old saying goes. And because actual fighting is expensive, most of making them tired devolves to training. Military forces train for every imaginable thing—from multiple imaginations. Many of these training sessions are devoted to the actual wartime or peacetime duties, but many are also general-purpose training: what to do if you're in this unusual situation and that happens. The unusual situation can often be tied back to primary duties, but it can take a convoluted chain of logic to get there.

In the meantime, teams of people who might have to work together tomorrow in a real life-and-death struggle find a way to get done what they have to do, while expending the least possible effort and time. Building teamwork and efficiency is the real goal of most exercises; skills are generally already there and are being improved rather than acquired.

Because of the ongoing training, when a military element is deployed in support of disaster relief, the leaders (typically officers) already have practice determining the important issues and deciding which things must be handled first and which can wait. The middle-managers (sergeants) already know how their teams will be organized and which teams have special skills. The teams themselves are already used to the job—both the working together part and the tackling the job without wasting (much) time whining about how unfair it is that we have to do this while they do that. The operational system springs forth because it is what these people always do.

Why don't more civilian organizations act like that? Some do: firefighters and SWAT teams, emergency room personnel, U.S. Forest Service Hot Shots, and so on. The reason they can respond is that they train and plan very much like the military—constantly and as realistically as possible. When they have to respond to the real thing, it's close enough to what they've already done, lots of times, that they can bridge the remaining gaps. Their responses, being practiced, are virtually automatic, and because they dissect their training after the exercise is finished, they know where other gotchas could have been, and they've thought of ways to handle those, as well.

You have neither the reason nor the budget (in time or money) to train at that level, but you don't really need to. When we pull together the common threads of all network disasters in Chapter 9, "Preparing for Disaster," and Chapter 10, "Returning from the Wilderness," we'll take a look at what training ought to do you the most good, most efficiently.

Attitude

All that training fosters an attitude that assumes success; a way will be found to solve the problem so we can go home and have a beer. The training has also, through experience, taught the importance of some details, making the leaders and middle managers attentive where they need to be. They, too, have learned through experience that they can do what is needed, when it needs to be done.

But the attitude that develops from training that is sufficient in quantity (covering breadth of possibilities) and quality (more intense than the real thing, if possible) is more than just "been there, we know how to do that." There develops as well, especially among the local-area units such as the Guard, a sense of responsibility for carrying out the mission to get things started on their way back to normal. No one expects to be able to get it all done in one weekend or, in the case of a major disaster like Hurricane Andrew, in two weeks. But it will get done, in a series of steps, and we have the job of doing this step now, so let's get it started. That attitude especially comes from the career NCOs, who have seen many situations and call on that real-world experience for perspective.

Prevalent among local units like the Guard or a local militia is a sense of serving the community. That matters in a number of little ways, such as going a little longer on a shift, walking security patrol one more block than required because it's all a part of the same neighborhood, and so on. That is the reason they are here, doing what they are doing; they didn't sign up to be a soldier so much as to serve their state or province, their community.

They are serving, not just working for a paycheck. Economists call this psychic income; it is a form of compensation, and you don't have to be a military or civic organization to engender it and get the extra from your people. You do have to give them a reason to offer it.

A Plan

Finally, military organizations bring with them not only a plan, but often a plan for this sort of situation. It doesn't exactly fit, but it's a starting point; it can serve as a guide. Even more important is the military's commitment to actually having a plan; because you need a plan to be sure everybody's working toward the same objective, if there isn't one, make one. That does not mean that, when the world is collapsing around your ears, a lieutenant will not help until she has drawn out a detailed plan, complete with all the appropriate appendices. It does mean that objectives will at least be sketched out, a desired sequence of events will be determined, and how we get the system we're in to start going in that direction is communicated to those who need to get things started.

Clausewitz warned that no plan survives contact with the enemy; the universe will not react quite the way we envision as we sit calmly in our office playing games of what-if-this or -that. So accept now that whatever plan you make will turn out to be somewhere between slightly-less-than-a-good-fit-for-the-circumstances and well-it's-a-start. Likewise, you will not be perfect in implementation. But, for lack of a better analogy, consider a professional athletic contest. Every coach/manager/trainer begins the game with a plan, one tailored to use his team's assets to their best advantage while minimizing the opponent's ability to do the same. Neither plan generally survives the first few minutes of the game intact; what does survive and helps the team through periods when the game may threaten to fall apart is the intent expressed in the plan. Otherwise, how could the almost-beaten opponent in any number of games manage the miraculous finish that wins in the final seconds (or even in overtime)? Examples of that abound, from Havlicek tipping away an inbound pass from the opponent and his teammate holding possession for the final few seconds to win the NBA Finals, to Manchester United scoring one goal in the 90th (unofficial) minute to tie Bayern Munich and then a second in stoppage time to win the Champions League Final, to the Chicago Bears scoring twice in the final few minutes (with a two-point conversion) to tie and then intercepting for a winning touchdown.

Plans kept carefully in the safe will not be implemented. That takes us back to training, which develops attitude. Organization, training, attitude, and a plan are not a secret recipe available only to military forces or professional sports teams. They are available to any organization, yours included. So how do you get them? You make them happen by taking things seriously and being interested in outcomes. You do not have to be personally involved; a useful exercise for your people is handling a hypothetical crisis with you not available for guidance. If they can handle things without you, they are competent as an organization, which is the best compliment to your capabilities as a manager and a leader.

What Are You Planning For?

That is not a rhetorical question, nor should the spoken emphasis be placed on the word "you." Plans are intended to help you in a certain situation, so start by defining the situations you must plan to handle. There is a continuum of warning: Some disasters arrive after there has been considerable warning (perhaps as much as 48 hours or more), and some arrive with (functionally) no warning at all. Most disasters' amount of warning actually lies somewhere between these two extremes.

Adequate Warning

This category is really limited to major weather events, such as hurricanes, typhoons, and cyclones, along with major winter storms. Every region of the globe is affected (these are not just tropical storms). In the North Atlantic, for instance, a plot of 10 Januaries' worth of storm tracks crisscrosses the ocean north of 45 degrees. Most storms pass north of the United Kingdom, over Scandinavia, then into Russia; a significant number, however, pass further south, even as far south as the Iberian peninsula.

In the Pacific, Australia's north coast was affected by 9 typhoons (from the Pacific Ocean) or cyclones (from the Indian Ocean) in 2000 and 2001. Further north, Singapore and the Malay Peninsula were approached by only 3 named storms since 1996; Taiwan, however, experienced 16 in the same period. Japan was hit by 17, but Korea by only 7. The economically booming southern area of China experienced 3 or 4 every year for the period, totaling 21. (Because reconstruction takes time and consumes capital, consider the effect this has on the infrastructure available to a business.)

The Indian subcontinent experiences 2 to 4 named storms every year, usually evenly divided between the east and west coasts, with the accompanying disruption of development, again, especially of infrastructure. Every few years, a north Indian Ocean cyclone veers onto the Arabian Peninsula, usually quite weakened by the time it arrives. In the southern Indian Ocean, the eastern coast of Africa and Madagascar experience 1 to 4 cyclones per year.

The United States experiences anywhere from 3 to 7 named storms every year. In the same 1996-2001 period cited, the United States was hit by 26 named storms. Approximately one-third were limited to the Texas/Louisiana Gulf Coast, a few less struck only Florida, and the rest tracked northeast up the Carolina and Virginia coast, with significant (though lesser) impact farther along the coast, all the way into the Canadian Maritime provinces. Not all made it to the category of major storms, but all had significant local impact with rain, flooding, and (along the coast) storm surge.

The quality of the warnings available for such storms varies, as does the timeliness. In part, that is because storm prediction is still not as exact as it could be. Part of the reason is that weather is chaotic—it exhibits sensitive dependence on initial conditions. We still do not have either the complete characterization required or the computational platform (hardware and software) to make use of it. When we really need to have the computational power readily available, and the algorithms to use on it, the impossible will become routine.

Even though we cannot predict a given storm with great reliability, there are trends to guide us.

Figure 5.1, for example, is a map from the United States National Weather Service showing the tracks of major hurricanes (Category 3 means sustained winds in excess of 111 mph/178 kph). These storms originated during years when the Sahel, the semi-arid grasslands south of the Sahara, had a relatively moist year. In dry Sahelian years, fewer storms approach the U.S. Atlantic coast. What you can see from the map is that roughly once every five years, on average, a major hurricane strikes the U.S. Atlantic coast.

If you expect to be in your property anywhere in this area for at least five years, you should have a plan that encompasses how to ensure your network keeps functioning in the event of a major hurricane.

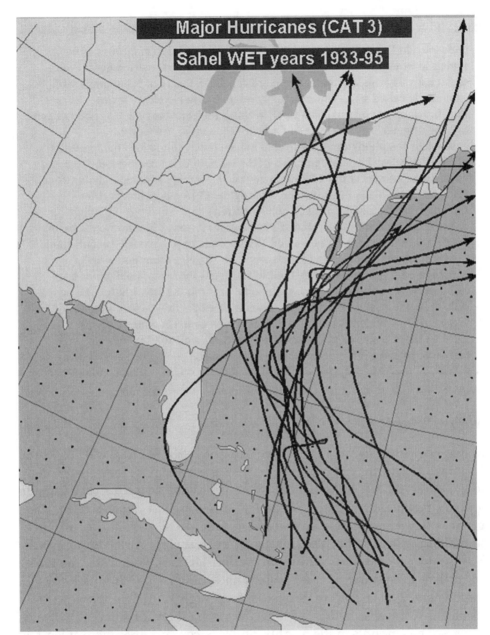

Figure 5.1 Major Atlantic hurricanes, Sahel wet years, 1933-95.
Courtesy of the U.S. National Weather Service.

> **POWER IS NOT ENOUGH**
>
> Code Excited Linear Predictive (CELP) is a method used to compress voice signals in order to use less bandwidth. The basic, uncompressed digital voice signal uses 64 kbps channels. These are called DS-0s; 24 of them make a DS-1 (commonly called a T-1, which actually refers to the physical medium specification), the original "fat pipe" in North America, and 32 of them make an E-1, the original "fat pipe" used in most of the rest of the world.
>
> Wireless telephony cannot afford to use so much bandwidth on one call; the needs of wireless have driven much of the work in voice compression. We are now able to compress the 64 kbps signal down to 8 kbps (commonly) or even 5.3 kbps through advances beyond the original CELP algorithm. Further compression is possible, but call quality suffers.
>
> The CELP algorithm by itself was not enough: It needed better hardware on which to run. CELP originally needed 125 seconds to process a 1-second sample of digital speech using a Cray supercomputer. The algorithm now runs in hardware on a specialized chip, the Digital Signal Processor (which has been responsible for much of Texas Instruments' profit in the late 1990s). The digital cellular phone nestled in your palm does with a DSP chip what a room-sized Cray could barely do in 1985, and it does it well over 125 times faster.
>
> When there is sufficient market pressure to find a solution, enough resources to make it happen do become available.

Not Just Hurricanes

Major storms, as mentioned earlier, are not just a tropical phenomenon. Major winter storms can paralyze a city, or even a region, for a surprisingly long period of time. In January 1998, a major winter storm struck Canada and the northeastern United States. Ice storms are not unusual in the area; in fact, they are common from eastern Ontario across Quebec to Newfoundland, a swatch that also cuts across upstate New York, Vermont, New Hampshire, and Maine. During the period January 5-10, 1998, the water equivalent in freezing rain, ice pellets, and snow ranged from a low of 73 millimeters/2.87 inches in Kingston to a high of 108 millimeters/4.25 inches in Cornwall; both are along the north shore of the St. Lawrence River, so the amounts in the neighboring U.S. communities were at least as great. This was roughly twice the greatest amount ever previously recorded in an ice storm in Canada. The duration of the storm also ran roughly twice as long as normal, delaying the beginning of repairs needed to the infrastructure.

This is the most densely populated and urbanized part of Canada and a major area of trade with the United States. Included in the storm path were the major cities of Ottawa, Montreal, and Quebec. Between cities, it is a heavily wooded as well as farming area. Not only were millions of trees lost (contributing to greatly increased fire hazard in subsequent years), but 120,000 kilometers/74,580 miles of power transmission lines were down, along with 30,000 wooden poles and 130 major transmission towers. That is a lot of infrastructure to repair. Three weeks after the storm's beginning, over 700,000 people were still without electricity in Canada.

While this storm was an unusual event that may be tied to the El Niño phenomenon, severe winter storms are a hazard across all of North America north of approximately 35 degrees north latitude. These may be Nor'easters in New England, ice storms in the mid-Atlantic to Great Lakes area, plains blizzards, and major systems striking the Pacific Northwest directly off the ocean. They do not always occur at the expected time; the U.S. Midwest was struck by a major winter storm in late March one year—a few days after all the snow removal equipment for one city had been put into storage. It took several days to dig out.

Major Storm Effects

While there are differences between the effects of tropical storms and major winter storms, there are also similarities. Both deposit enormous volumes of water on areas from which it cannot dissipate (either by melting or running off). Coastal zones, most subject to tropical storms, are typically flat and so cannot lose the accumulation quickly once the storm moves on. One example from June 2001 was Hurricane/Tropical Storm Allison: It dropped approximately 30 inches/762 millimeters of rain on the Houston vicinity. Areas that had never flooded before were under several feet of water.

Complicating the problem of runoff in coastal zones is the fact that sea level rises locally by several feet. The storm surge from Hurricane Hugo, a Category 4 hurricane in 1989, was 20 feet/6.15 meters; storm surges from even minor hurricanes can easily exceed 8 feet/2.46 meters. This rise in sea level is caused by the sustained powerful winds pushing up a wall of water, but while that wall is present, rivers cannot drain into the sea. In fact, seawater pushes inland and can easily contaminate local freshwater resources, including those used for urban water supplies.

Basement flooding is normal; lower floors are often flooded as well. Where is your networking equipment? Where is the incoming power supply (I should say supplies—you have redundant power, I hope, if you are running a large networking operation). If continuous network operation is important, you may well have a bank or two of batteries (heavy-duty car batteries on steroids, one of my colleagues calls them). Where is all this vital electrical equipment located with respect to the likely flooded area? Do you reasonably believe that water damage will not reach here? (Batteries and DC power lines, all together with computerized equipment that could short and spark in the presence of water—a not very happy thought.)

Thus, in a tropical storm you have time-compressed annual rainfall unable to drain away, and to complicate matters further, winds are (of course) extremely high. This eventually leads to power outages; even if local distribution is underground, transmission from the generating facility into the local area is not. Those lines are compromised, sometimes by shorts caused by objects being blown into them; even though the line still hangs properly from the tower, it is no longer capable of transmitting electricity.

You need electricity to power the pumps removing the water from your flooded basement and lower floors.

Perhaps you have backup generators. If they are at ground level, and if ground level is flooded (whether by storm surge or rain runoff hardly matters, does it?), the pumps are not only useless, they will emerge heavily damaged, to boot.

Finally, of course, there is damage from the wind. Actually, the wind, while breaking the glass in some windows, also causes glass loss via the objects it carries. Plywood sheets sell out when a hurricane approaches as people protect their homes (of course, you send your people home to do that, don't you?). But not only is there insufficient plywood to cover all the glass in an office building, there is neither the time nor the mounting surface (often) to nail or screw it down (screwing it down makes it both easier to remove and more readily reusable for something where the water damage is not a factor, like the next hurricane).

You may expect many of your windows to be blown out and large quantities of wind-blown rain to enter. This causes not only physical damage to the property (walls, floors, ceilings dropping from the weight of water above), but also damage to the equipment and furniture. Estimates by disaster recovery firms are that 60 to 80 percent of corporate data resides on local host hard drives. Do you back up those hard drives weekly? Are you prepared to remove all those desktop computers? To where? And with whom—you sent your people home to take care of their property, remember? Likewise, where are all the hardcopy files stored? Is that area reasonably protected from water and wind damage?

The damage done by hurricanes is well documented; we need not belabor it here. Likewise, the example of the January 1998 ice storm suffices to make you aware of the kind of damage possible from a major winter storm. A final point to consider as you plan for that is that these damages are not undone for weeks, sometimes months. In all likelihood some rebuilding will be required; you must expect to be unable to use your facilities in the damaged area for an extended period of time. And returning them to their prior functional use will be expensive.

Modest Warning

Major storms are not very common, but the cumulative effects of a series of lesser storms can cause a disaster as well, in the form of a flood or a landslide. Alternatively, a lack of storms can result in drought, which makes an area prone to fire. The denuded slope is then more prone to slippage when the rains return.

Lest you think these do not apply to business, the Seattle area has an average of approximately 13 landslides per year; not all are in residential areas. Landslides, in fact, occur in all 50 of the United States, in Puerto Rico and the U.S. Virgin Islands, and are even more common in the developing world where denudation of slopes is more likely to have occurred. As for the damage they can cause, even in the developed world, look at an instance not too long ago. In January 1982, an intense storm in the San Francisco Bay Area triggered 18,000 debris flows and landslides. Total direct damages were estimated at $123.4 million (in 2001 dollars). In addition, 930 lawsuits and claims in excess were filed against city and county governments for a total amount of $555 million (also 2001 dollars). Those are not household figures.

Likewise (apparently the Bay Area has better-documented hazards than many other areas), in October 1991, during an extreme drought, a vagrant had a small fire (in violation of emergency measures) on a steep hillside in a residential area of Oakland. Weather conditions were a firefighter's nightmare—65 mph/105 kph winds coming over the crest and down the slope, accompanied by record-high temperatures in the mid-90s F/~35 degrees C. The initial fire was quickly contained and extinguished, then closely monitored. A gust of wind blew a single hot ember into a tree just outside the previous burn area, and things exploded from there.

Assistance had difficulty arriving on the scene through the narrow streets, in part due to residents trying to evacuate and in part due to spectators arriving and browsing the area. The bodies of a police officer and a fire division chief were found with the bodies of those they were trying to help get out (a sobering reminder that heroics from such people did not begin on September 11, 2001).

Firefighters lacked water both due to the drought having drawn down area reservoirs and to residents wetting down their roofs and walls with garden hoses. Tanks could not be replenished because electricity needed to power the pumps had been knocked out as a side effect of the fire. Many mutual aid fire engine companies that responded to help Oakland could not attach to the fire hydrants; they used 2.5 inch/6.35 centimeter hose couplings, while Oakland used 3.0 inch/7.62 centimeter hose couplings.

Despite the technical incompatibilities, mutual aid support was extensive: 71 response teams of 5 engines each assisted, along with aircraft from hundreds of miles away making water and retardant drops. The overwhelming number of respondents clogged the radio frequencies, with communication made more problematical due to line-of-sight problems in the hilly terrain.

Costs of the so-called Tunnel Fire: 25 dead; 150 injured; 2,843 single-family homes destroyed, with 193 more damaged; 433 apartment units destroyed; 1,520 acres/615.4 hectares burned; fire perimeter of 5.25 miles/ 8.45 kilometers; $1.854 billion (2001 dollars) in damages. It could have easily begun in an area adjacent to light industry or offices.

Warning in these kinds of events is limited. Fires can move astonishingly quickly (ask anyone who has resided in Southern California more than five years for confirmation); landslides occur when the combination of moisture received and rising temperatures reaches a tipping point. In both cases, though, you will know that the underlying conditions are present (if you are paying any attention to the world outside your network). But in the case of flooding, there is lead-time, right? Not necessarily. Of course, there are flash floods (another one you may ask your Angelino friends about), but regular, major river floods need not take long to build, especially if you are in the area of the confluence of rivers, so that you receive the benefit (as it were) of the drainage from two basins, such as St. Louis, Missouri. Likewise, during the spring snowmelt, north-flowing rivers will suddenly accumulate enough ice to form a jam, and a flood will build very rapidly behind it.

Then there are man-made floods, perhaps most catastrophic to business, like the 1992 Chicago underground flood. Downtown Chicago has a maze of old tunnels from the late-nineteenth and early-twentieth centuries, when they were used to deliver coal and freight to office buildings. In more recent years, they have served as ready-made conduits for electrical and telephone/data cables and fibers. In April 1992 a construction crew inadvertently punched through the wall (which may have been weakened by seepage) and broached the channel of the Chicago River. The hole rapidly became the size of an automobile as water poured into the tunnels.

While crews frantically tried to seal the hole, the Loop shut down. At least nine buildings were flooded, and the power company shut down electricity in the entire district due to the danger of electrical fires from the exposure of the cabling to the floodwaters. The subway in the area also closed, and there was a shortage of buses with which to evacuate all the people. Water leaked into the Chicago Board of Trade building, so it shut down (the CBOT is the largest market in U.S. Treasury securities, among other financial contracts). Its sister exchange, the Chicago Mercantile Exchange, known for agricultural trading and more financials (forex futures, for one example) shut down gracefully at 10:45 A.M. lest it lose power catastrophically. Over a million contracts, valued at billions of dollars, are traded at these exchanges daily.

A similar incident occurred in downtown Dallas on Labor Day, September 2, 2000, when a construction crew installing a fiber-optic line punctured a 30 inch/76 centimeter water main in the heart of downtown. Severe flooding ensued, mostly on the surface, but flowing into basements, until the water could be cut off. Over 20 million gallons/72.7 million liters of water surged into downtown. At least $300,000 in damages was identified in the immediate aftermath; the contractor blamed faulty maps and markings on the streets, while the city offices involved, of course, blamed the crew for sloppy work.

Who was at fault might help you decide who to sue eventually, but it would matter not at all while the water poured into your facilities.

No Real Warning at All

Situations such as the flooding in Chicago and Dallas lead us to those disasters for which there is no useful warning. These can be sudden storms in addition to violent earth activities such as earthquakes and volcanic eruptions. A pair of tornados dropped into downtown Fort Worth, Texas, on a Tuesday evening in late March 2000. In fact, at 6:20 P.M., it was at the later end of the rush hour departures from downtown. The damage was so severe that the entire area was closed off until the following Monday. Seven buildings on the west end of downtown collapsed, and one 40-story building in the heart of downtown could not be salvaged. One reason to keep people out over the weekend was the removal of damaged glass from office towers; some windows weigh in excess of 200 pounds/91 kilograms and are an extreme hazard as they fall.

Tornado sirens gave a few minutes' warning.

On May 6, 1999, an even worse set of tornados (including at least one F5, the most powerful) struck Oklahoma City. There were at least 48 deaths, and direct damages were estimated to be approximately $1.25 billion (2001 dollars). Again, tornado sirens gave warning, based in part on a classic weather radar pattern. Surprisingly, there were no deaths at all in the 4-24 age group, an outcome with a probability under 0.1 percent. The reason? Oklahoma schools drill the children on taking cover during a tornado. Without the warnings *and the community's response to them* the death toll is estimated to have been 684.

Will your personnel respond? Do you?

Even less warning is usually available for an earthquake. With no warning, in 20 seconds at 5:46 A.M. on January 17, 1995, the city of Kobe, Japan, was devastated by a magnitude 7.2 earthquake. The damage extended

> **THE RICHTER SCALE**
>
> Earthquake intensity is most often expressed in a measurement along the Richter Scale, named after its developer, Dr. Charles F. Richter, of the California Institute of Technology. The scale was developed in 1935, and it is thoroughly embedded in the popular knowledge of earthquakes.
>
> What is not so well embedded is the fact that the scale is logarithmic: Each increase of 1.0 represents a tenfold increase in energy released, or power. The larger the amount of energy released during the event, the larger the amplitude of ground movement at a given distance.
>
> While there is no theoretical upper limit on the Richter Scale, you will rarely see a number greater than 8 in the units column. In fact, the following ranges may help you assess how much you should expect your facilities to be able to withstand (and, like the British Board of Trade Regulations for lifeboats, you may not want to assume that current building codes are necessarily sufficient):
>
> | | |
> |---|---|
> | < 3.5 | Quake not generally felt, though easily recorded |
> | 3.5-5.4 | Usually felt, but damage rare or insignificant (usually) |
> | ≤ 6.0 | Damage slight to well-designed/constructed buildings |
> | | Major damage to poorly designed/constructed buildings |
> | 6.1-6.9 | Destruction across a 100-kilometer/61-mile diameter |
> | 7.0-7.9 | Major earthquake. Serious damage over even larger areas. |
> | 8.0 + | Great earthquake. Serious damages in areas several hundred kilometers in diameter. |

over a 100-kilometer/61-mile radius and included metropolitan Kobe and Osaka. An estimated 5,500 people were killed, and the direct damage was estimated at $161.85 billion (2001 dollars). Business interruption and lost production are not included in those figures. All transportation arteries, road as well as rail, suffered major damage in the area, Japan's second most populous. Nine days after the quake, over 367,000 people were still without water.

Of course, all around the so-called Ring of Fire (the Pacific Ocean rim) the threat of earthquake and volcanism is high. In addition to the metropolitan areas of Japan, the most urbanized areas in the threat zone are the Seattle-Tacoma and Portland areas in the U.S. Pacific Northwest, the San Francisco Bay Area, and the Southern California metropolitan complex of the Los Angeles basin to San Diego in California, along with the major metropolitan area centered on Mexico City.

Probabilities of events may be estimated, though they do not give any guidance as to the likelihood of a particular event at a particular location in a given time window. Nonetheless, they do offer some guidance in construction and for operational planning. The United States Geological Survey estimates that there is a 70 percent chance of a magnitude 6.7-or-greater quake in the Bay Area by the year 2030 (Figure 5.2). 6.7 is the magnitude of the 1994 Northridge quake in the Los Angeles area, which killed 57 people and caused $22.5 billion (2001 dollars) in direct damages. The same study gave an 80 percent probability of a quake of at least 6.0 magnitude; depending on location, the damage from such an event would still be quite serious. For comparison, the earthquake at Loma Prieta in 1989, over 50 miles/80 kilometers away, measured 6.9 and caused severe damage in the Bay Area, including the famous collapse of the double-decker freeway that was scheduled for earthquake reinforcement.

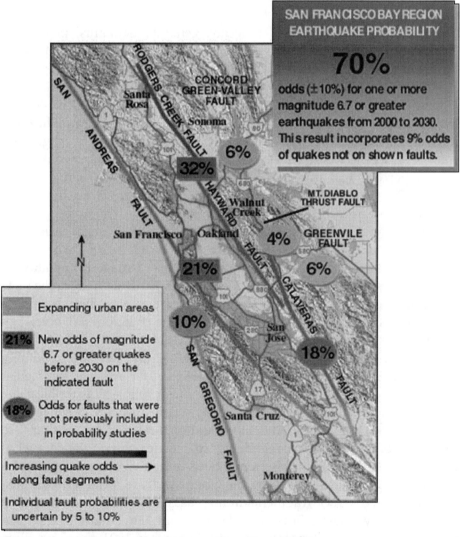

Figure 5.2 San Francisco Bay region earthquake probability.
Courtesy of the U.S. Geological Survey.

In examining Figure 5.3, it is worth noting that the buildup seen before major quakes of the past is not yet present; this is due to the enormous amount of strain relief from the 1906 quake. The USGS, however, does not expect this situation to last forever, or even for more than a few more years.

Likewise, models have been used to evaluate the likelihood of damage in the Los Angeles basin, a larger area geographically. The estimate is that somewhere in Southern California there will be a magnitude 7.0 earthquake roughly seven times each century. Much of this swath is mountainous forest, largely uninhabited, but a significant portion is not. Within the next 30 years, the probability is estimated to be 85 percent, though there is no way of predicting at this time where in southern California such a quake will occur, or even along which fault.

Earthquakes that occur under oceans generate tsunamis, great tidal waves. A tsunami is a pressure wave forced higher and higher into the air by the rising seafloor beneath. While tsunamis are less of a hazard than earthquakes, they have caused significant damage and loss of life. The National Geophysical Data Center of the U.S. National Oceanic and Atmospheric Administration maintains a database of natural hazards information. A search for tsunamis caused by earthquakes of magnitude 6.0 to 8.0 between 1959 and 2000 yielded 337 separate incidents, only 27 of which (8 percent) had any loss of life (and several of those were only one life lost). The overwhelming majority of tsunamis were along the Pacific Rim, with a few in the Mediterranean Sea.

In your planning, tsunamis are certainly exotic and may seem sexier than other events, but they are far less likely to occur, much less cause damage.

Also unlikely to occur, but more likely to cause damage in the affected areas, are volcanic eruptions. Likewise located primarily along tectonic plate boundaries (such as those around the Pacific Rim or the southern boundary of the Philippine plate, which is the southern edge of Indonesia), volcanoes generally give some warning, for some time. It is enough time, unfortunately, that people become inured to all the little rumblings and ash puffs, assuming that this time is more of the same. Mt. Saint Helens, for instance, rumbled at varying levels of intensity for two months before erupting catastrophically.

Lava is not the greatest danger to people and property, however. Far more dangerous are lahars, landslides, and pyroclastic flows. Lahars are mudflows composed of volcanic debris in a water slurry; they travel at high speed and therefore apply tremendous destructive force to obstacles in their path. Pyroclastic flows, on the other hand, are formed by the suspension of hot, dry rock fragments in hot gases. These move at extremely high speeds. Pyroclastic flows and lahars from Mt. Pinatubo are what buried Clark AB in the Republic of the Philippines in 1991; the flows filled the surrounding valleys with a volume estimated at 5.5 (± 0.5) cubic kilometers/~3.43 cubic miles of materials that hardened into something very like concrete. Based on warnings from geologists of an imminent eruption, the Philippine government evacuated 60,000 people from the surrounding valleys, and the PAF and USAF evacuated more than 18,000 military members and families from Clark AB.

People may have complained about evacuating at the time; no one seemed embarrassed after the eruption.

Mt. Pinatubo's eruption is described by the U.S. Geological Survey as 10 times larger than that of Mt. Saint Helens. The latter eruption has a well-documented sequence of events, beginning with a magnitude 5.1 earthquake that struck in the early morning; within 15 to 20 seconds the entire north side of the mountain slid away in the largest landslide in recorded history (Mt. Pinatubo had no significant landslides; not all volcanoes do). The huge loss of mass depressurized the underlying magma chambers, which exploded through the sliding debris. (The quantity may be appreciated by the fact that, during the preceding eight weeks of rumbling, the lava dome on the north flank had pushed outward by more than 100 meters.) The material became a pyroclastic surge, traveling at 500 kph/300 mph down the northern valley. As material fell out, the flow decelerated. Several people escaping on the western side were able to outrace the pyroclastic flow in their automobiles, driving at 100-160 kph/65-100 mph. No one worried about oncoming traffic.

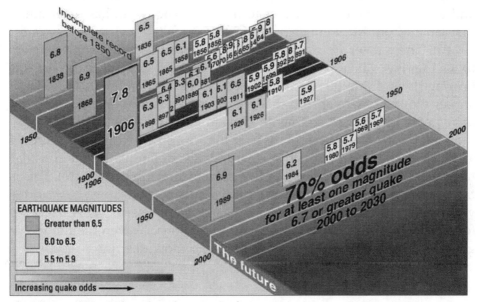

Figure 5.3 Historical earthquake magnitudes.
Courtesy of the U.S. Geological Survey.

The effects spread downstream into major rivers. Thirty-one ships in the Columbia River were stranded in upstream ports after the volume of material flowing down reduced the river depth from 40 feet/12.31 meters to 14 feet/4.31 meters. Reservoirs used for power generation were likewise partially filled, and they could no longer hold the volumes of water they once could. (The extreme power prices in California in the summer of 2000 resulted partly from less hydropower available in the Pacific Northwest as a result of drought; the Mt. Saint Helens eruption has not been directly implicated, but reservoir capacity was reduced.)

While Mt. Saint Helens was certainly dramatic, unlike Mt. Pinatubo, it did not affect many people and affected even fewer businesses. Unfortunately, it is hardly a unique event, even in the history of the area. Mt. Rainier, the beautiful and serene presence southeast of Seattle and Tacoma (and about half as far away as Mt. Saint Helens), did the same thing approximately 5,600 years ago. In fact, a series of lahars has raced down from Mt. Rainier in the intervening time; at least 6 and possibly as many as 13 have inundated the lower valleys. The last one, about 500 years ago, reached a depth of at least 50 meters/162.5 feet in Puyallup, then spread out across the valley floor to a depth of 10 meters/32.5 feet. This is the area between Seattle and Tacoma. A much smaller landslide broke free from the same area on the mountain in the early 1900s.

Worth remembering is that Mt. Rainier is not only the highest mountain in the area (at 14,410 feet/4,393 meters), it has the greatest load of glacier ice of any volcano in the conterminous United States. Melted ice would add enormous amounts of water to the flow. The USGS has determined that lahars from Mt. Rainier (including that melted ice) are the greatest hazard. The danger from lahars and subsequent sedimentation extends down the river valleys all the way to Puget Sound. Several of the streams flowing down from the massif have power generation dams; these lakes are far too shallow to absorb the load of even a minor eruption, adding dam breaches and the resulting flooding to the damage equation.

Finally, there is one form of disaster that can strike with no warning at all. It is not, however, exactly a natural disaster (unless you consider human carelessness and/or stupidity an elemental force of nature). This is the human screw-up that catastrophically cuts you off from the world—you and several thousand other users. That kind of disaster we will reserve for Chapter 8, "Unnatural Disasters (Unintentional)," even though we've glimpsed it already in the man-made floods of downtowns.

It's a Scary World, Isn't It?

And yet we manage to survive, don't we? Natural disasters are disasters in part because they are infrequent, not an everyday occurrence. The problems occur because we are not prepared for the events that have unfolded. We are in the wrong place, we have the wrong (or insufficient) equipment, we didn't believe the early indicators, and now what do we do?

Rudely put, if we had our kit in one bag, we would not be in this mess.

That is your job, isn't it? You are reading this book because you are responsible for keeping your network going in the manner to which your business has become accustomed, regardless of whether or not all hell has broken loose in the skies above your head or in the earth beneath your feet. The natural disasters you might face are determined readily enough. Local governments almost always have some sort of emergency management function; start there. Better yet, decide who is going to do the research (you may really not have the time, but the job must be done—delegate it) and have him or her start there. Make clear that you need to know what a detective or a journalist always asks—what, where, when?

You also need to know a few more things, like how often? When was the last occurrence? What is government likely to be able to do/not do (what do I need to be ready to handle myself, whether or not I should have to)? National government agencies are another source of information; I have cited a number of U.S. ones in the course of this chapter, and several of their Web sites (current as of this writing) are in Appendix A, "References." If you are responsible for a global network, your local people should know the government resources to check locally. You can also search on the topic "natural disaster" and find many summaries available on the Internet; without government sponsorship, their information depends on the quality of their sources but is also less likely to be tainted by political window-dressing of any sort.

When you know the natural hazards your location will realistically face (as opposed to the exotic and newsworthy ones we all read about and Hollywood makes exciting), you can begin to plan how your sites will cope. You must take the lead when it comes to designing, implementing, and practicing the plans to handle natural disasters. Doing that is the subject of the next chapter.

CHAPTER 6

The Best-Laid Plans

*But, Mousie, thou art no thy lane, In proving foresight may be vain;
The best-laid schemes o' mice an' men Gang aft agley,
An' lea' e us nought but grief an' pain, For promis'd joy!*
Robert Burns

Burns' farmer's plough may have turned over the Mousie's nest and ruined her plans indeed, but the poet would not have argued from that outcome that it is useless to have a plan. Just don't expect things to fall out perfectly in the end.

For that matter, there are plenty of sources for disaster planning for a company, along with business continuity planning (and there are some links in Appendix A, "References"). Our focus is narrower, however: This book (and so this chapter) is about your network continuing to perform its business function despite a disaster. In the case of a natural or unnatural disaster, you may be able to continue from the present location; you may have to operate from an alternate. Either way, the point is that, for the business to continue, the nervous system must continue to function. That's our job.

Three Main Points

USAF Squadron Officer's School almost had a fetish about Three Main Points (students even chanted about them, "...See how they run"). Humans do seem to be fond of breaking things into threes (even Julius Caesar said "Gaul is divided into three parts."); stories have beginnings, middles, and ends; love stories are more interesting when they involve a triangle.

Likewise, there really are three principle areas of concern when planning for a business to continue despite some sort of disaster: operational continuity, taking care of the people, and taking care of the network assets. Even though taking care of the people is usually more the concern of the corporate disaster plan, we will spend a little time on it because people are a special asset, perhaps the most important network asset you have.

They are the asset that makes all the others useful.

Operational Continuity

Redundancy is expensive. Not being redundant is even more expensive. Which would your Board of Directors prefer—to pay now or pay later, potentially a great deal more?

You are responsible for the networks that deliver information to and from your customers, whether they are internal or external to your corporation. Your network has some areas that are not that critical and some areas without which the network will not be able to deliver information. The latter must not be single points of failure in order for you to have operational continuity (nor should they be cut off from the rest of the network due to any single failure).

That does not necessarily mean that you must have duplicate equipment and facilities for all your critical connection centers. It does mean that there must be a redundant means for performing the same business function. Secondary systems are often of lesser quality, but as long as they are good enough for long enough, they are sufficient.

That, in turn, means two things. First, this is a judgment call as to what constitutes both "good enough" and "long enough." The answer for a large public university is different from that for a national newspaper with a large electronic subscriber base, and that in turn is different from the answer for a carrier or service provider. Second, you will have to make compromises between what you would like to be able to do and what you can afford to do. When you make that choice (or, actually, when you present choices to those who will decide for the corporation), you must quantify the risks and costs as well as you can, and you must be blunt that these are estimates of what you could foresee. They are no more perfectly predictive than a weather forecast or a stock forecast or a particular date of birth (barring a planned Caesarian delivery).

As noted previously, network continuity means operational continuity, but that refers to the network's capability and performance rather than the continuing operation of a particular piece of equipment. Operational continuity requires that you identify those network capabilities that are critical and ensure a backup means of delivering them to your business. Your backup may be a contracted service with another party, but some extra due diligence is required if you choose this route. If you are protecting yourself from a hurricane striking Charleston, South Carolina, and your site provides storage area networking in the southeastern United States, a backup supplier of storage whose communications are as vulnerable to the hurricane as yours are is no help.

Twenty Questions

To prepare for operational continuity, you must answer some questions and report those results to your management. Tucking the answers away in a report filed and forgotten serves no one.

There are rather more than 20 questions here, but their purpose is the same as those of the puzzle-solving game—to progressively define the possible answers. They are a start to your planning, but they are by no means a complete list; you will think of other questions, perhaps triggered by these, perhaps in the middle of the night (write them down; the odds are heavily against your remembering them in the morning). Some of these questions need no elaboration, while others have baggage, implications that come along and should be addressed. (All the questions are listed together in Appendix B, "Questions to Ask Yourself.")

Above all, think about these before you answer them. The first answer that comes to mind may well be accurate; it is less likely to be complete.

- What capabilities are critical?
- What facilities provide those capabilities?

The first question seems simple, but it is not necessarily. For starters, you should define what exactly your network's capabilities are before you decide which of those are critical. To apply an economic analysis, the network creates value by enabling the transport of information from where it has less value to where it has more value; it serves an intermediation function. Which capabilities are critical to that, and which are merely useful?

Name-to-IP-address mappings stored on a server have no intrinsic utility, but name-resolution accessibility by hosts with browsers is necessary for any large network to function. Note that I did not say the name-resolution service had to be provided by your network, only that such a service must be available. In answering the first question, define the capabilities that must be present, not the ones you must provide.

The second question begins to address the latter; the rest of the answer is deciding which you can outsource in times of need and which you must provide yourself.

- Will the entire capability need to be replaced (is partial function possible in some situations)?
  - Can the missing portion be provided from another location, or must the functions be collocated to work properly?

This reminds you not to assume that things must work in a given way; it is possible that a failure of one aspect of an important location need not require a complete assumption of all that location's functions by another location. As a small-scale example, perhaps you have a network backbone with large-capacity switches. At one location, one blade of a switch fails; you do not replace the switch (unless this blade is the last straw in an ongoing battle with your vendor, and, even then, you probably keep things going from their normal location as best you can until a replacement is tested and configured and you have cut over traffic to your satisfaction). There are some functions, though, that do need to be collocated; if one portion of that goes down, the set should switch to the redundant operation, not just the failed portion.

- Can other, existing facilities provide those functions now?
  - Is this one-for-one, one-for-many, one-for-all redundancy?
  - Can they support more than one failure (or do you wish to assume that is too unlikely)?
  - Is it a collective redundancy (workload distribution of one failed location among all others)?
- Does everyone know who will get what?
- Does the redundant location have the information needed to assume the function?
  - Can it get the information in a timely fashion in a crisis?

With flexibility, and possibly some rearranging, you need not purchase anything to have sufficient ("good enough, long enough") redundancy. You must evaluate what effect that choice has on your network. Is there, for instance, one location that is capable of serving as an adequate backup to several? That location then becomes more critical than its ordinary functions by themselves would warrant. If your functions are well distributed, can essentially any location back up any other? How will they know who should act first?

Suppose you have "Follow the Sun" coverage for customer support: a center in Memphis, one in Kuala Lumpur, and one in Florence. Each is capable of backing up either of the other two. Do you have a rule, such as the next westernmost site picks up the function if able (so that Kuala Lumpur backs up Memphis, which backs up Florence, which backs up Kuala Lumpur)? How will a given site know that its backup function is being activated? Do they all access (functionally) the same database, so that when Florence is responding for Kuala Lumpur, the person assisting the customer has all the relevant history in front of him or her? Will the customer and the representative be able to communicate (all responders must speak passable English, say, but does the customer necessarily expect to have to deal in English if he's used to talking to Malaysia)?

- Does everyone know how to activate the redundancy?
    - Is there a procedure established?
        - Is a particular person's/position's authorization required?
        - Does that person/position have a backup for vacation, incommunicado, and so on?
        - Does the backup person know when he or she is "it"? How?
    - Has the procedure ever been tested?
        - Were the problems fixed?
    - Was it retested to check the fixes?
        - How long ago was that?
        - What has changed since then?

Have you ever tested the redundancy activation? Would a customer have found the experience acceptable? Your opinion, or my opinion, or that of anyone except the customer does not matter. If the redundancy does not deliver to your customers—internal or external—it does not deliver. Internal customers you may be able to browbeat by outgunning them in corporate fratricide ("My VP's bigger than yours"), but that neither changes the fact that your system did not deliver nor does it give the internal customer any reason to assist or support you in any way in the future.

External customers won't be around longer than they have to be, and they won't be inclined to listen.

- Is there a critical person?
  - Is there one key person who can always make things work?
  - Is he or she always available?
  - If he or she left the company, would the critical knowledge leave, too?

And, of course, there is always someone who can make things work when no one else can. That someone knows the little quirks of the network, the minor things that solve major problems (at least, temporarily—and temporarily may be "good enough, long enough"). Does your redundancy system need that person in order to work? How embarrassing ... especially when he or she leaves for a better job. For all the military's dumbed-down reputation for procedures, you rarely find a situation in which for want of a nail, a kingdom was lost. Standardized, documented procedures not only mean you are less vulnerable to the loss of a solo genius (who might be out with chicken pox, having never had it as a child), they also mean that you have records of what works. And when you are not in the middle of a crisis, thinking about those fix-it procedures may tell you something about your network's behavior.

- What is the probable cost if you do not provide redundancy?
  - What are the direct costs (Service Level Agreement penalties, for example)?
  - What are the indirect costs (lost contract renewals, lost new business, litigation, and so on)?

What happens if you've thought about redundancy and you can't afford it—you just can't get there from here on your budget? Then you have a responsibility to let the corporation know the risk it runs. You need to document what will happen in the event each of your critical capabilities is lost. Your documentation should specify, in as much and as precise detail as you can truthfully provide, what will be the consequences for your network's customers. Provide your best estimate of the likelihood of each capability loss (and if it's not a great estimate, but it's the only one you have, say so). If you know there will be performance penalties or other direct costs, estimate them. If you have reason to believe there will be contract nonrenewals, litigation, or more, document those outcomes.

- What alternative means of providing the capability can you acquire?
  - How much does each cost (initial and lifetime costs)?
  - Do they enhance ordinary operations in any way to offset some of that cost?
  - Can they be made to do so?

It may be possible to provide the capabilities in a manner that fulfills other needs, thereby offsetting some of the cost. It is likely that the initial cost will have to be borne, but it is possible that the lifetime cost can be offset by sufficient benefits to make the purchase worthwhile, depending on your corporation's time horizon. Of course, if the corporation is cash-flow-challenged (to be polite), perhaps a standby arrangement, fully paid only when it is activated, might provide sufficient backup. The fee to maintain the right to activate may be affordable compared to the loss you would take, weighted by event probability.

- How long can you operate in this mode?
  - What are the operational effects of going "too long"?
  - What are the personnel effects?
  - What will your customers think?
  - What will you tell them?

Part of your presentation must be how long things can reliably operate in redundant mode, and the limiting factor there is more likely to be people than equipment. It could also be the expense of the unusual communications services you had to activate to operate from the alternative location. That alone may help you justify the additional expense required to be at least somewhat redundant with internal systems.

The last little question arises only as a follower to the higher-level questions, but it can make or break your redundancy plans. Spend some time lurking on any mailing list of network operators/users, and you will find profound skepticism of promises made of return to operation by carriers or service providers whose connectivity has gone down. Such skepticism is rooted in experience.

If you lie, or even waffle, to your customers about the duration of your outage, they will never forget. And they will not hesitate to repeat the story to anyone who asks about your performance. How much marketing budget will it take to overcome that? More than your corporation can afford, especially when you are shedding customers and their revenues. If you don't know how long you will be down, give an estimate, but be sure your representatives say it is only an estimate. Honesty really is the best policy—and that requires you to be honest with those who are facing the customers and their questions. They are personnel affected by this outage, too, and their professionalism will be priceless.

- When operating from redundant facilities, what becomes the new critical capability? (Do *not* assume it remains the same.)

Once you have things roughed out, take another look at how the network would operate in this mode. The changes you are operating with will place new stresses on your system, and you must consider these secondary effects as well. You may well find that the redundant-mode operations have created a new weakness or vulnerability that must be addressed. This is an iterative process, and you will have to choose an endpoint to the iterations; specify it and state why this is the endpoint (even if it is that things are now too difficult to predict because so much has changed that network performance simply cannot be estimated).

You are not doing this as a fundament-covering exercise. You have a requirement to deliver a business function, and if you cannot meet that requirement for reasons beyond your control, it is incumbent on you to report that to those who will hold you responsible. They may be able to change the environment and thus enable you to deliver, or they may not. Either way, they are warned, and doing so in writing serves two purposes. First, reading it will gain more of their attention than a conversation in the hall. Second, in the process of writing it, you will have to organize the information, which improves the presentation.

Finally, as you review the situation, bear in mind that a wonderfully designed redundancy that is not activated is both expensive and, unfortunately, too common. Human error, in the form of poor judgment, lack of information or initiative, or lack of authority, can render all your preparations moot. That is the reason behind all the checklists and repeats you hear, and not just in the military. Pilots repeat their clearances back to air traffic control; people are required to speak oaths in their own voices, not just agree that something sounded good. What was said, what must or will be done, is specified in a format that all must be conscious of.

A Few More Questions

Once you have some idea of how you will be redundant and how you will activate that redundancy, you must consider a new series of questions.

- Who must be notified internally?
  - Who will make that notification?
  - Who is the backup for making and receiving notification if the primary is not available?

Internal notification, oddly enough, may be more problematic than external notification. Do you have a manager with a "shoot-the-messenger" approach? If so, that person will be notified late, if at all. If he or she occupies a position involved in the redundancy activation or implementation process, you have a real problem.

- Who must be notified externally?
  - Partners.
  - Customers.
  - Vendor support (be first in line for recovery by being first to notify...).
  - Do both the primary and the backup notifier have those telephone numbers in a portable device?

This relates to the previous question about what you will tell your customers, but it takes things further. There are many others with whom your network interacts, and you need to let them know how to access your information. Perhaps you provide information via an extranet to customers; you may need to give them the backup IP address or server name so they can point their server to it (ideally, you do this in advance, so that your interruption is functionally invisible to them because the backup address is already loaded as a secondary destination address in their network).

- Do emergency services know anything about your facilities?
  - Hazardous materials like battery backup banks and legacy fire suppressants.
  - Presence and location of DC power sources (inverters, for example).
  - Locations of substantial electrical equipment.
  - Locations where the last personnel out of the building would probably be.
    - Their probable egress routes.

You should consult with your local emergency responders (fire/police/ambulance) as to what other information would be helpful for them to have in advance. They may also be willing to do a walkthrough of your facility; if so, they may be able to offer suggestions that improve their ability to respond and could minimize your damage in the event of a disaster. In this area, they are the local experts, and applying an ounce of prevention that they recommend may save you many pounds of cure after the fact, especially if it means you can recover more equipment and need to replace less. If you solicit their recommendations, either follow them or document why you did not. If there is litigation later, you will need to be able to justify your actions or lack of them.

Finally, by specifying procedures as thoroughly in advance as possible, you minimize the range of things that must be "good enough, long enough."

Getting the People Out

There are basically two categories of people for you to worry about: those on-site and those not on-site. Because the on-site personnel are your responsibility, you want to minimize their numbers to keep the problem manageable; that implies that you want to prevent more of your people (or anyone else's, for that matter, except emergency response personnel) from coming in.

From a sheer cold-hearted calculation, you want to keep as many of your people not only alive, but well, as possible. When you reconstitute this location, you will need their familiarity with how things were set up and working before to minimize the time it takes to become operational again. In addition, during the interregnum, they may be able to assist (remotely) those who are backing them up when the latter must troubleshoot the portion of the network the former normally manage.

Off-Site

Because prevention is much easier and cheaper than fixing a problem, we'll look briefly at handling the off-site people first. Having decided that the local operation must be shut down due to the disaster, you will want to notify those who might try to come in to work anyway, whether out of ignorance, disregard for their own safety, an exaggerated sense of their importance, company loyalty, fear of losing their job, an attempt to impress their supervisor, or any other reason that makes sense to them. Depending on the normal length of their commute (which in metropolitan areas can exceed 90 minutes), you need to make the notification available at least 2 hours in advance to wave them off. Your decision time must then be backed up from that by however long it takes to begin the dissemination.

You also need a dissemination means that is known, will reliably be available, and will, in fact, reliably disseminate the information. For instance, you post a sign by the exit door reminding people that, in case of bad weather, they should turn to a certain radio station. That requires assurance to you that the station will be on the air and will broadcast your notification (for a fee or as a public service?), and that your notification will be repeated frequently enough (among how many other notices, not to mention the regular programming?) that you can be confident no one should miss it.

A more common alternative is a recorded message at a given phone number, advising people to come in normally, come in later, or not come in at all. This is more reliable because you control the message (you don't depend on the radio station getting it right when its primary business is something else). The lead time for decision making remains the same, except now you must call in and record the voice mail message that your callers will retrieve. In addition to reminding people of the phone number, you could choose one within your number block that is easy to remember, like XXX-9675 (XXX-WORK), or set up a branch off your base dial-in number ("for information on facility operations, press 7 now").

For smaller, critical-manning operations, you may wish to set up a telephone tree. This requires discipline to maintain (someone has to be sure that it is current) as well as to operate (knowing who to call, who to call if that person is not available, etc.). Telephone trees are used by the military, but responsiveness is required there, and serious consequences exist for a chain not being completed in a timely fashion. The recall is also tested periodically; that is less likely to work in a purely civilian environment, though the same groups who adopt military-style training and organization also tend to adopt these practices and make them work, for the same reasons. If speed is of the essence, you can set up the early branches of the tree to call two people each instead of only one; however, in that area, if one of the two is missed, the caller must pick up both next-layer people or you risk losing an entire branch.

On-Site

Personnel who are at your facility when you decide to transfer operations and evacuate are your responsibility. Part of your overall business disaster plan should include how to make the decision (who decides, the fallback decision maker, decision criteria) and where people should go. Most people who are present at the site have no great operational responsibilities and are immediately free to go. You should have a marshalling area where they gather, and you can document who got out (and who is not present); this both simplifies the job of the emergency responders and reduces your legal exposure later if you can establish that supervisors were held responsible for their people (and it may even establish that a missing person was known to be outside the building at a certain time). Once you know who is accounted for, then release them or send them to a known safe area (pre-arranged), as per your disaster response plan.

How much of your building's physical security depends on a properly functioning network? Consider a facility with a default physical lockdown to prevent theft (of anything and everything from artworks to information to money itself); if the network shuts down, doors default to a locked position, with the exception of those required by fire codes to be able to "crash open." When those doors are opened, do they trigger alarms? If you shut off the alarms, how do you know that no opportunistic person is taking advantage of the confusion and the alarm status to steal what is normally protected?

You may need to retain a minimum of people to operate your local electronic nervous system; they and physical security are likely to be the last ones out. That should not be a surprise to anyone. In the interest of protecting the company and supervisors (presumably including you) from litigation, there should be written notification that this is part of the job and they should be required to acknowledge that they understand that.

A copy should be kept securely off-site, of course.

Network Assets

Networking hardware, especially WAN hardware, is surprisingly tolerant to harsh conditions. As we'll note later, several WAN and telephony switches were discovered still working well past their expected survival time in the disaster site from the World Trade Center attack. The hardware may well survive, at least for a while (flooding tends to be a bit harsher on the hardware, of course, than a major winter storm).

What about the configurations?

If they are stored in RAM, when power is lost, they are gone. If they are stored on a hard drive (a named file or startup configuration/committed provisioning), is the stored copy the same as the one that was running when things went bad? Not necessarily. It is not uncommon to make some changes to a network configuration, keep them in RAM so that they can be activated, but not make them a part of the permanent configuration until they have proved themselves stable under normal loads for a certain period of time. This leaves the possibility of a relatively easy rollback to a known reliable configuration if things go bad with the changes (our best-laid plans, again).

Even if they were stored on the hard drive, what if it is not salvageable? Perhaps there was a fire, and the building's fire suppression worked well. Unfortunately, water damage is not good for electronics. Insurance (or financing) will pay for replacement hardware, but who carries a map of the network in their heads?

If nothing else, a paper copy of the network structure (physical and logical topologies) should be stored at your alternate operating location. There are two good reasons to have it there: first, they need it to operate as much of your portion of the network as possible while your location is not available; second, it is unlikely to perish in the same disaster.

I'm sure you'll keep it current.

Even better would be a soft copy of all configuration files (they can typically be saved as text files), stored on a server (with backups elsewhere), using a naming convention that includes the device name and the date of the configuration file. From this storage, the alternate location would be able to retrieve the files and know what they are dealing with. Likewise, a soft copy of the physical and logical topologies would be usefully stored there, as well. This is easier to keep current, as well as ensuring rapid retrievability if needed. (Note: These files must be at least as thoroughly protected from intruders as the most sensitive corporate financial information or intellectual property, or even more thoroughly protected—they are the keys to the kingdom for a hacker.)

Network elements are typically password protected (you have changed the vendor defaults, haven't you?); does the alternate location have access to the operator account and password to get in and modify configurations, if necessary? A standard account and password used on all elements means you needn't keep a file somewhere (which is a risk), but it exposes you to a greater risk—a hacker who compromises one node can have them all. At some point, you have to trust your people; this is your chance to choose the point.

Beyond choosing good real estate, there is not a great deal you can do to protect your assets in the event of a natural disaster. Do consider the property aspects when the opportunity arises. If you are currently in year two of an eight-year lease, there is not much you can do, but if you own the property, you can consider improvements, trading off their cost against the protection they could offer, weighted by the best estimate of a given disaster's probability. Of course, your best chance to apply disaster prevention considerations is before you choose the property, bearing in mind that you are likely to pay a premium for safety.

This premium must also be considered in conjunction with other costs associated with the different location, such as greater/lesser connectivity expense, greater/lesser utility expense, transportation issues for your employees, and more. Asset protection must be evaluated financially, as well as operationally, as part of a system of costs and benefits. This is unlikely to be your decision alone; quantifying probability-weighted costs and benefits will give your input more weight in the process, however. If you are not familiar with establishing probability-weighted costs, it will be covered in Chapter 11, "The Business Case."

These questions will give you the framework within which you must create a plan for network continuity. Most plans assume you have the chance to think, but real events tend to happen faster than we are prepared for; unless you practice plan activation (could you even find the disaster response plan, right now, if you needed it?), you are likely to be behind the power curve after approximately 30 seconds.

No-notice events, such as sudden violent storms and earthquakes, where the warning has become part of the background noise and has passed out of our attention, and sudden fires inside the building will require your people to activate the plan. That raises an interesting question: Do they know the plan exists? Here's another question: Could *they* find it in a hurry? If they passed off functional responsibilities to another location so that they could evacuate, would the other location know what to do? Would they understand what was going on and the need for prompt action so that the others could let go?

No-notice activations will stress your organization's implementation of the plan; it should include that possibility. When you practice the activation of the plan (a subject we will return to in Chapter 9, "Preparing for Disaster"), you should include no-notice situations as much as possible. You may find that aspects of your plan require notifications that may not occur or formal transfers of authority that cannot happen because one party is not available. Evaluating the exercises will probably cause you to revisit the questions. This will be an ongoing process, not a check-off-that-box-once-and-never-look-again exercise.

If you need to understand why it is important to practice, here is a simplified situation in which there is a plan, or, more accurately, pieces of a plan, but the questions have not really been addressed. It shows.

Example: Data Services

Consider a hypothetical data services provider, CDG, Inc. A provider of high-touch voice and data services to small and medium-sized businesses, CDG began in the Kansas City metropolitan area and has gradually expanded to Minneapolis-St. Paul, Chicago, and Denver. It currently offers VoIP and Virtual Private Networks (VPNs), and it is preparing to provide a security services package that will include network monitoring, filtering, and firewall operation. CDG currently offers a package deal arrangement with Secured Storage, Inc. (also entirely hypothetical), which provides off-site data storage for clients in the same metro areas via CDG's connections.

Many of CDG's customers take advantage of Secured's storage program, which does a master backup once per week, with incremental copies daily on the other four (or five, or six, as required) days. Secured has staggered the day for master backups among its customers, though Monday tends to be a little heavier than the other weekdays and the weekends are lighter. Backups are automated, and Secured has staggered them, as well, to minimize the bandwidth required from CDG; they run from midnight to 1:45 A.M., then skip the potential daylight time change block, resuming at 3:15 A.M. and continuing until approximately 4:45 A.M.. If Secured obtains two or three more data-intensive customers, it will need more bandwidth to get all the backups done in the allotted timeframe. Secured uses data links from CDG to replicate its data between metro locations.

For redundancy, each CDG operating location has dual connectivity to the Secured location in the same area, via two different physical paths, though with the same local carrier. Each CDG location connects to two other CDG locations, forming a virtual ring, as shown in Figure 6.1. CDG also has a VPN on this ring, used for its own corporate business, including exchanging soft copies of all network configuration data—topologies, addressing, and startup/committed configuration files as part of database replication.

By bundling the connectivity requirements of many smaller businesses into one account, CDG is able to secure favorable rates from its carrier. Much of CDG's appeal to its clients is that, in addition to offering VPNs and VoIP at prices better than they can obtain individually, CDG offers reliability. First, it has its logical ring. If the carrier's links should fail, CDG has a contingency frame-relay service with a second carrier, for which it pays a fee to maintain the right to activate it with anywhere from 4 to 12 hours' notice; the rate charged on activation depends in part on the amount of lead time CDG provides the carrier. Activation requires the authorization of CDG's Chief Technology Officer, Shu-Ming "Shelly" Guo, one of the company's founders. The connection to the second carrier is preconfigured at each CDG location; CDG need only load that file and have the carrier activate its end of the circuit, and the backup is on line.

Based on this level of reliability and the assurance of a person on duty all the time so a business with a connectivity problem can troubleshoot whenever it needs to, CDG has just begun supplying service to a home improvement chain, along with three other new customers with locations in Kansas City and the Twin Cities. If these customers are still pleased at the end of six months, their business will be the lever that enables CDG to expand to Memphis and Fort Worth, perhaps even further. The newly configured connections came up like clockwork last Thursday and have been stable ever since, which bodes well for the expansion.

CDG has a business continuity plan for each of its locations, as well as a master corporate plan. If Kansas City is unable to perform corporate functions, they will hand them off to Denver; if contact should ever be completely lost, Denver will assume them after 24 hours. Site operational responsibilities shift clockwise around the ring (Chicago backs up the Twin Cities, and so on).

The last week of January each year, Shelly Guo and her family take a skiing vacation in Taos, New Mexico. Leaving during the usual "January thaw" is part of the fun for her, as any accumulated snow is likely to melt

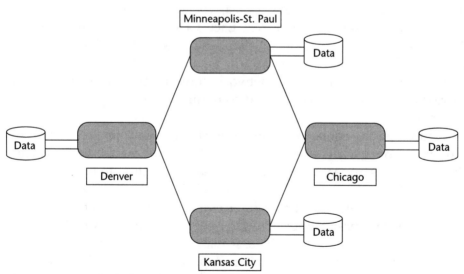

Figure 6.1 CDG logical topology.

and leave things sloppy with no real prospect of spring for another six weeks. She prefers the neatness of deep snow and has lobbied her cofounders to move corporate headquarters to Denver so that she could ski more often. Because they have no profound attachment to Kansas City, they are considering it. They accept her assurance that it will make no technical difference (they have marketing and financial backgrounds, not technical; the three made a balanced team).

Two days after Shelly's annual departure, a cold rain that becomes sleet and will eventually turn to snow arrives. This is not unusual in Kansas City, though it can leave driving something of an adventure in the many hilly areas of the city. This storm does not move on through due to another system blocking its path further east. Fortunately, the Monday day shift of three network operators was already on duty when the weather turned bad. Unfortunately, the city has issued an appeal to all motorists to stay off the roads until conditions improve; a second storm is moving in and may well combine with the remains of the current one and make conditions extremely hazardous.

CDG sends most personnel home as soon as the word gets out of the worsening weather. One of the network operators is released, and the others will hand off to the two evening shift personnel when they are able to get in; Kansas City does keep major thoroughfares well plowed and sanded or salted.

Temperatures fall more slowly than expected, and the precipitation continues to be freezing rain instead of snow. Not only do the streets remain extremely hazardous, power lines begin to go down, often due to tree limbs crashing across them as the weight of accumulated ice becomes too great. The downed power lines make the streets even more hazardous, and the second-shift personnel give up trying to come in after they are more than six hours late. Third-shift personnel have checked with the city and been advised that, unless they are required for emergency services (hospital/ambulance employees, telco/electrical repair, etc.), they should stay home.

The two lucky network operators have learned that the choices in the candy machine don't make much of a dinner. They divide the remainder of the night into two-hour blocks and plan to take turns napping and monitoring. That is thin coverage for the data backups, but because they don't normally work the night shift, they don't know how much detail they will need to track. One begins looking for the binder with ordinary procedures while the other tries to rig a sort of bed from cushions from the guest chairs in a nearby office.

At 11:42 P.M., the power fails. Emergency lights required under the fire code come on, illuminating the rooms with their harsh but effective light. After a heart-stopping moment, the on-duty operator realizes the UPS has kicked in and his PC starts to reboot.

Every other piece of network equipment does the same.

The second operator returned immediately, not having really gotten to sleep and aroused by the power loss followed by emergency lights. Operating from the surge of adrenaline, they make a quick pass around the room, checking each piece of networking equipment at the physical level (all power lights on, all blade indicator lights or link lights correct, etc.).

With relieved sighs, they settle back at the operator's station and call the power company to learn how long it may be before power comes back on. The estimate is unknown due to the many outages being worked; theirs is in the queue, but there is no way to predict a duration at this time.

One operator goes back to the log entry when they came on shift, now over 16 hours ago. They had dutifully checked and recorded the configuration file running on each piece of equipment, a task usually considered a nuisance. Just as a double-check, they will duplicate the review; they really should have done it when the second shift actually began, even though they had made no changes. Because they are now into third shift, they ought to do it just to fill that square.

Six of the eight files are not the same.

The six devices whose filenames differ are those that were reconfigured with circuits for the new customers, with data coming in from multiple locations in the metro area. The two operators recognize that; what they can't find, in their surprise, low blood sugar, and fatigue, is the file on each hard drive, the new topology diagram, or the new configuration settings. While one searches, the other calls Denver to see if someone there can locate a copy to send back. Denver asks how long the UPS battery will last—something they hadn't checked. Needless to say, the UPS batteries were never intended to keep the center running for several hours, only long enough to transfer operations to Denver and let them activate the backup circuits.

Denver takes over operations, but it cannot create (in just a few minutes) connections to Kansas City's customers that do not pass through the Kansas City node. The net result of this is that the Kansas City customers are without connectivity to the Internet, voice service, or backup—and that is all of CDG's Kansas City customers once the battery runs out. Those customers were single-homed: All their connectivity ran through CDG's Kansas City location.

DEFINE "UNINTERRUPTIBLE"

All uninterruptible power supply (UPS) devices are not created equal. There are four basic types, with variations or combinations available. The four types are standby power, ferroresonant standby, line-interactive, and online power. They are listed in order of their responsiveness as well as expensiveness. You will see some resources list only three types, lumping ferroresonant and other line-interactive types together; though the name used for this type varies. I prefer to break them out here for clarity.

Standby power supplies have an internal feed to charge a battery, while the bulk of the incoming power passes through surge suppression and filtering to smooth out any unevenness in the power supply. At the power egress is a transfer switch; if the main power fails, the transfer switch connects to the battery and power flow resumes.

Note that verb: *resumes.* There is a brief delay during the loss detection and switch process. If the delay (which should be included in the UPS performance characteristics, listed as transfer or switch time) is significantly less than your equipment's power supply hold time, there is no problem. This sort of UPS is suitable for PCs and smaller, less critical devices.

Ferroresonant standby power supplies offer an improvement in the way the transfer is made. Both the main wall power and the battery (charged from the wall power, as before) are connected to a transformer, from which the power egresses to your equipment. The transformer performs the line filtering and conditioning; in addition, its internal field acts as a buffer, supplying that brief moment of power during the switch from the wall to the battery.

Ferroresonant power supplies are larger and can handle greater loads, making them suitable for some of the smaller networking applications; however, they are not as efficient as other, newer designs, and so they are beginning to lose market share.

The third popular design is the line-interactive UPS, which runs power to the battery via an inverter/converter (remember, batteries are DC and require an inverter from ordinary AC power); the egress comes off the inverter/converter. When power from the wall stops, the flow *to* the battery becomes a flow *from* the battery.

While this has the same lag time problem as the standby power design, it does come in larger sizes. If you can live with the lag, which really is quite small (check your equipment's power requirements—telephony switches are very sensitive, but data equipment is less so), this can be a suitable solution.

Finally, there is online UPS (some call it "true" UPS, especially its manufacturers). In terms of circuit design, online and standby UPS systems are superficially the same—both have power coming in from the wall, where a portion runs through the battery to a transfer switch to the egress, and another route goes through filtering and conditioning devices to the transfer switch to the egress.

(continues)

> **DEFINE "UNINTERRUPTIBLE"** *(Continued)*
>
> Note the order I listed the circuits in, though. The main, everyday, ordinary power, in fact, is running through the battery, while the filtered and conditioned wall output backs that up. The reason is simple—because power is coming from the battery all the time, there is no transfer delay—zero—when wall power drops. Therefore, the supply of power is truly uninterrupted (hence the marketing of this as "true" UPS). The "backup" of filtered wall power is available just in case something goes wrong with the battery-based circuit (an inverter failure perhaps—possible, though very unlikely). If you must switch to the wall circuit, there will be a small delay; the difference is that the delay is far less likely to happen. Also, by filtering routinely through the battery, there is no direct link from any burps and bumps in the line power. While filtering and conditioning really does remove all but the very worst of that, some can remain; charging the battery and then drawing from it removes that from your list of worries.
>
> Online UPSs are more expensive both because they are built for larger power throughput and because building an inverter that must operate continuously is more expensive than making one that must operate only long enough to gracefully shut down some PCs. Unfortunately, because the incoming power is converted twice (AC to DC to AC), there is an efficiency problem as well as significant heat to dissipate (exacerbated by the fact that this type of UPS is used where large quantities of power are needed).
>
> You have to examine your equipment's power requirements and decide how fast your conversion must be. Then you can determine which type of UPS you need. This is not an area to skimp on, either. With standby or line-interactive/ferroresonant systems, if you have a UPS too small for the load, it may simply fail to kick in due to the overload, start and then fail itself, or start and fail by blowing a fuse or tripping a circuit breaker. Warning lights on the UPS may well have indicated the potential overload, even when power is coming off the grid—but someone has to look at the system to see that. No network monitoring function will report it.
>
> An online UPS may simply fail to work when overloaded, which at least lets you know at a time when you can do something about it (because you draw power through it routinely).
>
> One last point: Be conservative about what must run off the UPS. Printers (and laser printers are power hogs) can wait. Even some servers could conceivably wait—it depends on the content they are serving and to whom (if their clients are not on the UPS, why do they need to be up?). Use the UPS power only for that equipment that must keep going.

Like the *Titanic*, where no one imagined a long and sliding collision down the side of the ship, no one planned for the entire Kansas City node to be out of service. In fact, as we examine CDG's topology again, we can see that each city, by having one location and running all connectivity through it, is a single point of failure for all that city's customers.

Several problems are apparent in CDG's situation; let's take it apart and find where things failed, where they didn't but could have, and what CDG did right. When you evaluate your network plans in light of the natural phenomena with which your operations could collide, it is important to discuss all three categories. Don't forget to point out where things should go right—knowing they've already won part of the battle encourages people to take on the rest. Acting as though nothing is right discourages them from even trying.

Lessons Actually Learned

This example is greatly simplified in terms of network size and scale; the reality of such an operation would be more complex. The point, though, was to create a situation in which certain weaknesses would be exposed by a collision with natural forces. Among them were a flawed topology, insufficient facilities, poor configuration control, and inadequate equipment.

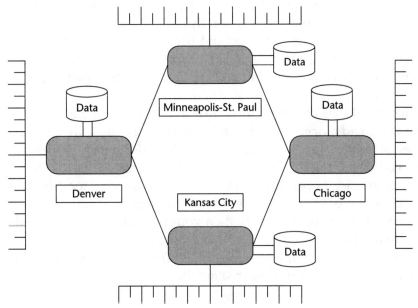

Figure 6.2 CDG expanded topology.

Topology

Each city has a dangerous topology; there may be redundant links to the Secured Storage server farm, and even redundant links from customers to CDG, but each CDG Network Operations Center (NOC) is a single point of failure. This is a situation where Murphy's Law will, sooner or later, apply; the single point of failure will fail, and probably not for the reasons you expect and for which you make contingency plans. In Chapter 8, "Unnatural Disasters (Unintentional)," we'll see a few situations where the one place where all the pieces came together failed, for reasons no one anticipated.

Because we don't have perfect anticipation (those who do make their living by taking bets from those who don't—in the stock market, for instance), we must design the system so that there is no single point of failure. If you are unable to do that, you must be honest, even painfully honest, with your customers about what that point is and what you believe to be the likelihood that it could fail (but make sure they understand that this is an estimate, not a guarantee).

Facilities

In an adverse weather situation, it is possible that your 24/7/365 operations center will be manned by people who cannot leave to go home. Is there anything present to support their extended tenure? Is there food? (Tap water may not be fashionable, but it generally is safe.) Candy machines are not suitable food; microwave popcorn and heatable noodle cups require (surprise) electricity. If the weather conditions that keep people there are likely to involve loss of power, plan so that their food is something that does them good.

How much should the company provide the continuous operations center? Decide that in advance, communicate it to everyone, and there will be no surprises, at least in terms of what is available under which conditions. And when you decide, bear in mind how important continuity of operations *in this location* is; if operations can be shifted to another location and this one could disappear from the topology (not just be unattended—it could be completely nonoperational), then your personnel support needs are greatly simplified.

From experience, I recommend against military field rations for your emergency food. Camping supplies are *much* better.

Configuration Control

In this situation, new configurations had been created, they had been stable under operational conditions, but due to insufficient elapsed time or human error they were never saved to the startup/committed files that the equipment reverted to on reboot. This equipment should not need to be rebooted, and no conditions should arise that cause an inadvertent reboot ... but things happen anyway. They should at least have been saved to a filename in a standard directory location on the device's hard drive; that file could be reloaded after a reboot.

Likewise, the startup/committed files are exchanged with other sites on CDG's VPN; operating files should be exchanged, as well. Without configuration updates being saved and exchanged, if another site took over control of this location's equipment, their understanding of the detailed topology would be based on an outdated and inaccurate configuration and might not include support of new customers.

As it turned out, given that this location was a single point of failure, the configurations being accurately stored here and at other sites (in the database) became irrelevant when the UPS's battery ran out, which is our next failing.

The Right Tools for the Job

The UPS had a delay in kicking in, which means it was not an online UPS. For an operations center committed to 24/7/365 duty, that is probably a mistake. It is not definitely a mistake because it is possible that the delay time of a UPS is well inside the power disruption tolerance of the equipment in question, but examine the issue very carefully. Use of an online UPS would have prevented the reboots.

Likewise, how long the UPS can support the equipment's power needs must be examined carefully. In this case, it was sufficient to hand over operations to another location. The fact that topology issues made that useless means that this location (and every other CDG location until the topology problem is rectified) must be able to operate as long as it takes for local power to be restored. As an alternative, it might be necessary to keep things going only until generators can be engaged to provide power.

Those, however, bring another set of performance criteria to manage, as well as periodic testing (to maintain engine performance and to ensure power output meets quantitative and qualitative parameters) and the monitoring of consumables (such as fuel and lubricants). After all, if the UPS kicks in and bridges power until the operators can step outside and start the generator, but the generator's spark plugs are corroded, what have you gained?

This was a hardware lesson, but it could just as easily have been software, or wetware (personnel). Do you have the pieces in place right now that can handle the job in adverse circumstances?

Lessons Potentially Learned

With Shelly Guo out of town, could CDG have activated its backup connectivity (on-call frame relay) if its carrier had gone down? Who was Shelly's backup? Was there no backup, in which case they had to call her on her vacation? Would they be able to reach her on the plane? On the ski slope? Or (the hypothetical I always use with my manager), suppose she got run over by a truck? Any way that she is out of contact is a time delay that circuit activation might not be able to bear.

CDG is used for writing backups essentially during the entire first shift—what happens if the power glitch occurs then? Is the backup software that Secured uses tolerant of interruptions, or does the job have to be restarted from the beginning? What will be the effect on bandwidth requirements in that case (because Secured has staggered the start times of the client backup routines to optimize their bandwidth requirements)?

Related to that, suppose the power outage had only lasted 35-45 minutes. CDG could have been back online for some of the backups, but which devices did it need to get up first? While presumably every client of Secured is backing up every night after a working day, on a Sunday, some of them may not be writing at all, and every day writers will effectively be in a queue: Their calls will come in on different connections at different times, for different durations. It would behoove CDG's operations center to know the sequence so that, if there is a general problem, its staff can activate connections in the order they are likely to be needed, rather than wasting time at 12:48 on a connection that won't be used until 4:12 when another is about to be tried.

Kudos

Yes, there actually are some positives in this scenario, in two broad areas: redundancy and personnel. CDG had taken care to have redundant connectivity between its center and Secured Storage, and its logical ring ensured multiple connectivity paths to every other site in its network. CDG had also contracted for responsive backup connectivity between its sites (with a variable fee, depending on how much lead time it could give the frame-relay provider). It even intended to have redundant power; its failure was one of implementation (given its actual topology) rather than missing the need for redundancy.

The two lucky operators made a considerable effort to keep CDG going, even when things just kept getting worse. They stayed on duty rather than going home and leaving the facility unattended. Network equipment, properly configured, really does not need babysitting, though an on-duty troubleshooter is a good thing to have when reliability is your market differentiator. With out-of-band connectivity, the network center could easily have been managed by another location (out-of-band would make the management less dependent on the primary carrier between locations).

The operators also called another site for help when they realized they had a configuration problem. It is amazing how many times people do not take this step; we all hate to admit to anyone that we need help, that we made a mistake. These operators put getting things back together ahead of their pride.

And the other location (Denver) tried to help. With no pressure on the Denver staff and without the fatigue and missed meals, they were thinking more clearly and realized the UPS duration was going to be a critical factor; they didn't keep that to themselves in some sort of petty one-upsmanship between sites. They took over operations, for as long as Kansas City's power lasted, to try to do what they could.

Extending the Example

As stated earlier, CDG is a simplified example. Suppose you are responsible for the network of a large corporation, with multiple major locations, perhaps like those in Figure 6.3.

This depicts a North American topology, assuming that your company takes advantage of NAFTA. Each location has redundant paths to every other location as depicted on this map. Suppose you have multiple locations in Chicago and Toronto and every other city here. Even if you have multiple routes over multiple carriers using different physical paths, if all your traffic for a city or region goes through one network center to your backbone, everything that connects through that one center is isolated if it goes down.

That is the same problem as CDG, but now it's a bigger company. And it's yours. And it's no longer so hypothetical.

If you are responsible for network survival and performance for your company, it's worth getting a few of your top people, the ones who really know your network, together. Sit down with a topology, maybe just a rough drawing on a whiteboard. Later, run the same exercise with some of your sharp, but less experienced people. Without the presence of their seniors, they may offer some strange possibilities. Some of those may be quite plausible, but would be seen only by someone new enough to not yet have preconceptions of how things ought to fail. Each time, start assuming total loss of use of a given site; can the information serviced by that site still reach everyone who needs it? Can those who need access to resources inside the blocked site still get them or their equivalent from elsewhere on the network?

If not, you have CDG's problem, just on a larger scale.

The collision of the *Titanic* was not what we normally call a natural disaster because no great expanse of territory or dense concentration of people was grievously altered when a natural event intersected the activities of those people. For the White Star Line, though, it was certainly a disaster that resulted from a natural event badly handled by the people who collided with it (made far worse by the planning of the people back home). The capital asset was a total loss, and fatalities reached 68 percent of the passengers and crew.

Could your network survive the equivalent in the form of a natural event that your firm handled badly through your planning and real-time choice of actions?

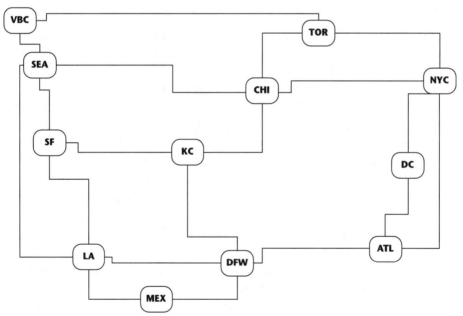

Figure 6.3 Continental topology.

That is where you should focus your natural disaster planning—handling the kind of natural event that is likely to occur in the location where you operate. Assume that Murphy not only will be present, but will be an active participant in the game. In Chapter 9, "Preparing for Disaster" and Chapter 10, "Returning from the Wilderness," we will look at ways to set up your network operations to improve your chances of network survival. The lessons you learn from the little exercise in this chapter will be an input.

Unnatural Disasters (Intentional)

You can discover what your enemy fears most by observing the means he uses to frighten you.
Eric Hoffer

Terrorists have a very high success rate in meeting their immediate goals, but a terrible one in achieving any form of long-term success. That may well stem from their misunderstanding of how societies, as opposed to groups of people, work. But if you're in the middle of one of those short-term events, knowing that is small comfort.

This is a chilling topic, and descriptions of events or procedures may offend some readers. I do not apologize for this. Terror is literally a matter of life and death; your preparations to handle it must never discount that, in any sense.

Terror, as a method of warfare, is present throughout recorded history. The first change, in the modern era, was an escalation in the degree of violence used. Even then, this was a function of increasing technology as the Industrial Revolution gave us more capabilities, and the advances of science (especially chemistry and physics) led to a better understanding of what made bombs explode and how to further improve weapons technology. The terrorist's goal still tended to be a political change, and it was important for the terrorist not to alienate the general population, whom he claimed to serve and whose support he needed for concealment.

The United States government uses this definition of terrorism: the unlawful use of force or violence against persons or property to intimidate or coerce a government, the civilian population, or any segment thereof, in furtherance of political or social objectives. The dictionary defines terrorism as "the systematic use of terror, especially as a means of persuasion." Of course, we have to look up "terror." When we do, we find the first definition is "a state of intense fear," followed by other definitions, and culminating with "violence ... committed by groups in order to intimidate a population or government into granting their demands." Both of these definitions match the historical record, but recent changes in the use and users of terror may cause our understanding of the term to evolve. The Bremer Commission (named after its Chairman, Ambassador L. Paul Bremer), in its report submitted in June 2000, noted that the issuing of demands for political changes or the release of imprisoned terrorist comrades was no longer the norm; terrorism was being committed by fewer groups, but they did so for the sole goal of inflicting mass casualties on a target population for religious or ideological reasons. The commission also did not flinch from the fact that terrorists attacked American targets more than those of any other country.

Whether there is a political demand of any sort, terrorism still focuses on the use of premeditated, extreme (by the victims' standards) violence directed against symbolic targets, targets chosen for their emotional impact more than for any true military value. Why choose such targets? Two reasons: First, they are soft; they are not only generally not well-defended, they are almost impossible to truly defend. (Example: Think about the resources it would take to truly defend a hospital against a truck bomb; after all, ambulances and cars ferrying emergency patients drive right up to a wide-open door. And if this hospital is too well defended, then we can move on to another one, or a school, or a set of children's athletic fields—the concession stand usually has a substantial number of targets—that is, people—around it.) The second reason is that the terrorist wants the emotional impact of unsuspecting victims, people going about their ordinary lives. The message sent by the successful attack on such a target is that you are not safe ... you will never be safe because I can reach out and kill you at my whim, on my schedule, for my reasons. You are not a civilian; you and everyone you care about are my enemy and therefore my targets.

The terrorist's definition of an enemy is not determined according to the rules of war, especially those developed in the European-North American culture. The modern terrorist's enemy is (often) that culture itself; therefore, everyone and everything associated with it are legitimate targets. There simply is no distinction between combatants and noncombatants, military and civilian, the government and the populace, men and women and children. In fact, those victims least likely to be considered targets are ideal because they demonstrate the reach—both the power and the will to use it—of the terrorist. They also emphasize the differences between them and us, playing into the cultural conflict. Therefore, women and children are better targets than men, the populace is better than the government, civilians and noncombatants are better than the military and other combatants (such as the police).

Targets should have symbolic value, collateral damage is a good thing, and the shock and horror of an unprepared victim are best. That makes a North American/Western European business location, preferably a trophy property, with Very Important People and lots of female staff, a wonderful physical target. That trophy property may well be your headquarters, the place where all your information flows, and, quite possibly, a major physical node on your network as well as its major delivery point.

One final point to be aware of: There are actually very few terrorist organizations, and their member population is not large. We can never really predict where they will choose to strike, though, because the number of possible targets that would make sense to them is so large. This leads to our needing to be prepared everywhere; universal preparation is hard to maintain because "everywhere" is never a target. Very few actual attacks occur, and there really are far, far more hoaxes (see Figure 7.1).

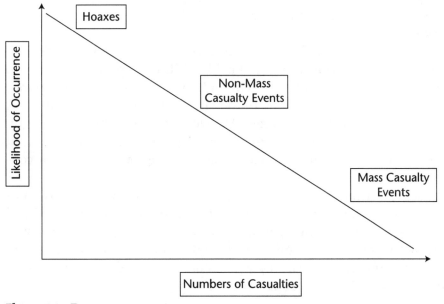

Figure 7.1 Terror events spectrum.

Because of this, we tend to relax our precautions and get sloppy in our measures to defend ourselves as time passes. This tends to result in an attack, when it does occur, that costs unnecessary damage and, especially, far too many lives.

Physical Attacks

Of course, you may not be the direct target. Instead, the physical targeting of a terrorist may have collateral damage that affects your network.

For instance, targeting a trophy property like New York's World Trade Center disrupted many, many networks because of the major switching centers located in or near that property (due to the volume of traffic in the area). Either way, you may lose physical connectivity to a major portion of your network, and you may lose additional critical assets, as well. Where, for instance, is your redundant storage? Some firms have it on a different subnet, on a different floor, but in the same building for good physical asset control. That's fine, until the entire building is a casualty.

Likewise, some firms in the World Trade Center had their redundant storage in the other tower; no one imagined a catastrophe so enormous it would take out both towers. Because of that, no one extrapolated what would happen if both towers came down (the construction engineers used the term *pancaked* for the type of collapse in place; it could have been worse, had they toppled sideways)—the surrounding buildings, which had been relatively fine until then, suffered enormous damage, which led to further collapses.

Or we could consider the Pentagon: a series of concentric rings connected by corridors. When a substantial blast hits a building, where do the fireball and concussion wave go? Through everything nearby, of course, but material objects absorb some of the energy, while corridors give that energy an easy path to keep going. An overhead photo of the Pentagon published in the days immediately following the attack showed what appeared to be soot or other smoke damage tracing out the corridor connections.

Depending on the building's fire suppression system, the smoke and heat that travel down the corridors may be enough to set off overhead sprinklers some distance away from the physical damage; how will overhead water affect your systems? (Even if you don't have overhead sprinklers in your equipment area, if they go off on the floor above, are you sure no water will come down through your ceiling?)

Bombs

Bombs have been a preferred terror weapon virtually since the beginning of modern terrorism, which most historians place in the late nineteenth century. Some form of bombing occurred in a strong majority of the attacks that have taken place in recent years; from 1994 to 1999, for instance, most attacks involving U.S. citizens or assets took place in Latin America, and the overwhelming majority were bombings. A total of 613 Americans were wounded; 64 were killed. On a global basis, businesses were the preferred target in this period (1,589 attacks versus 107 against government sites, 44 against military locations, and 584 against all other targets; likewise, American businesses had 133 casualties, compared to 9 each for military and diplomatic sites, 7 in government facilities, and 26 in all other types of facilities).

Bombs are not terribly difficult to construct (and the learning curve ensures there are no stupid bombers), they do not require large numbers of people to create or use (the fewer involved the better, from a security standpoint), and they have great effect. The heat and blast wave, perhaps carrying additional debris and/or shrapnel, does extensive damage. The function of a bomb is really nothing more complex than to cause a large, essentially instantaneous, transfer of energy into objects ill prepared to receive it. The innovation, if you wish to call it that, of September 11, 2001, was that the delivery vehicle itself was the bomb—a high-speed jet with a transcontinental fuel load delivered an enormous amount of energy to a specific target.

If the goal is property destruction, the bomb manufacture will emphasize heat and blast wave (concussion) effects. If the goal is to kill people, some of the energy will be absorbed by an extra payload—nails and other metal intended to be carried by the blast wave as high-energy projectiles. Both will cause physical destruction to buildings and their contents; the difference is in degree and in the additional casualties inflicted by the shrapnel payload.

Electromagnetic Pulse

At the time of this writing, an electromagnetic pulse (EMP) bomb has not been used, to my knowledge, by terrorists. The possibility has been discussed among network operators, however, and there really is no inexpensive protection against it. Where a conventional bomb causes damage to structures via the physical transmission of energy, an EMP bomb would cause damage to all electronic equipment—and only that equipment—within the range of its pulse.

For those not familiar with it, an electromagnetic pulse is a surge of electromagnetic force that transfers some of its energy to any electricity-carrying medium it passes through. If the pulse is large (which is the point), the circuitry can be melted, or even vaporized. If not, the electric circuit becomes a carrier for the pulse, which is no longer dissipating according to the inverse-square law. It flows down the circuit until it meets an interface incapable of carrying the load, at which point damage occurs (the cable may well be able to carry more power than the components of the circuit board to which it connects, for instance). EMP is generated, among other ways, by the detonation of a nuclear warhead. This need not be an exo-atmospheric detonation, though, if one of those were to occur, the effects would range over a continental-scale landmass (and network survival would become secondary to personal survival).

Most empirical data regarding EMP was developed before the bans on nuclear testing, so it is dated; some information is publicly available regarding the effects of a both an above-surface and a surface nuclear detonation. In the latter, a torus (whose dimensions would depend on the actual warhead size, type, and elevation at detonation) surrounding the blast point would be affected by EMP; this is beyond the area of shock and blast damage. Thus, a terrorist detonation of a nuclear device in a city might well have electronic effects that carry some distance away, ruining data and telephony switches, routers, and more.

It is not actually necessary to detonate a nuclear device to create an EMP. Any large and catastrophic discharge of electromagnetic energy would also act as an EMP. The principles are fundamental physics, and the implementation is electronics; testing of such weapons may have been done by the western militaries as early as the mid- to late 1980s. I found technical descriptions of such devices in a simple Internet search. Unfortunately, a like-minded terrorist could do the same.

Sabotage

Sabotage derives from the practice of disgruntled people throwing objects into the machinery during the early days of the Industrial Revolution in protest of the loss of work and disruption of their lives (the objects were, indeed, sometimes their sabots, wooden shoes, which would seriously damage the equipment). Sabotage can be a form of network attack, which we will address later in this chapter, but it can also be physical.

Consider a highly automated manufacturing facility; the automation is almost certainly controlled via software delivered by the network. The network can be the means of delivering the instructions that commit sabotage in such a facility. As an example of what could happen, recall the disaster at Bhopal, India.

Union Carbide built, in association with a local partner, a pesticide manufacturing plant outside the city of Bhopal. Squatters and slums developed in the area surrounding the plant (it was originally a little distance away from the population center). On the night of December 3, 1984, a cloud of methyl isocyanate (MIC) gas leaked from the plant and blew over a portion of the settlements. The Madhya Pradesh state government report cited approximately 3,800 deaths, 40 cases of total disability, and 2,680 persons permanently partially disabled. Other reports, usually provided by parties attempting to increase the payout from Union Carbide, cite much larger numbers in all categories. Union Carbide went through years of litigation, including continuing litigation after paying a $470 million final settlement in 1989 ($610.3 million in 2001 dollars).

The plant had a number of processes that had to be modified in order to create the MIC gas; physical sabotage was alleged but never proven. Likewise, a whole series of software-controlled safety systems had to be bypassed to run the infamous tests at Chernobyl (tests to which the on-site operators objected and were overruled). Does your company have a network that delivers operating instructions to physical processes? Could someone use the network to alter those processes?

It would not necessarily be a catastrophic event, though that is the likely outcome of a terror attack of this sort. And it has the same "elegance" as the use of the delivery vehicle as the weapon on September 11: It uses the culture's achievement as the means of inflicting great damage.

CBR Attacks

Chemical, biological, and/or radiological attacks are less likely, but they are not out of the question. They are also very hard to defend against. I've carried a gas mask in a pouch on my hip, but it had simple training filters in it. The wartime filters—the ones that could have actually defended me against a Warsaw Pact gas attack—were in secured storage (as they should have been—how many little things do we all lose track of?). Even so, at the first hint of the presence of gas, we were trained to don, clear, and then warn: pull the mask out of the pouch and unfold it, pull it over the head (taking off my glasses, a small delay), pull the elastic straps tight, press my hands tight over the intake zones, and exhale—with the breath I had left at this point because you do *not* inhale during any of this—hard enough to blow out any gas that was trapped inside by the donning process. Then, and only then, do you warn other people. The reason? If you delay to warn others, you may become incapacitated and the warning would not even be given—and you will have made a useless sacrifice.

To give troops an incentive, one exercise was held in a room filled with tear gas; the troops noticed how it was no problem at all, with the mask properly on. You learned to appreciate just how good the mask was when it became your turn to remove the mask, recite name, rank, and service number, and then don and clear your mask. I could not recite those fast enough to avoid getting some of the gas in my throat (via nose or mouth hardly matters); I was able to don and clear, but I coughed and gagged for some time, and my clothes reeked of it. That was simple tear gas (though a heavy amount for training purposes). How well could your people respond, without gas masks, to a simple tear gas attack?

Gas is difficult to do well; though heavier than air, finely atomized it will spread on the air currents, and so it dissipates rapidly unless confined to a space (such as a tight, weather-proofed modern building). A well-ventilated building will, unfortunately, disseminate the gas rapidly. Gas also needs to be at the appropriate concentration to do the level of damage desired (though if the desire is to kill, there is a great deal of room for error on the high side). There are four basic forms of gas agents: skin (blister) agents, choking agents, blood agents, and nerve agents.

Skin agents include tear gas, which is essentially an irritant to the mucous membranes (in the correct concentration; at higher levels it can be quite damaging). Other skin agents cause blistering from chemical burning of the skin, which is extremely painful. Such agents do severe damage if they contact the eyes or get into the respiratory system. They are, however, usually intended to debilitate rather than to kill.

Choking agents are intended to attack the respiratory system, leading to breathing difficulties, the filling of the lungs with fluid (edema—if severe enough, drowning in one's own body fluids), pneumonia, and so on. They are debilitating to fatal, depending on the degree of exposure.

Blood agents attack the circulatory system, usually penetrating the skin and then using the very system they attack to disseminate the weapon through the body. Blood agents prevent the transfer of oxygen from the blood to the cells, suffocating the victim at a cellular level; cyanide acts as a blood agent. There is no antidote available for a blood agent attack; these agents are intended to kill.

Nerve agents, such as sarin, are essentially a network attack on the human body. They disrupt the information flow to the brain and the command flow from it to the various body systems. They are chemically closely related to insecticides, and the manufacture of the latter can be diverted to the former with relatively little effort. They are absorbed through the skin and the eyes and inhaled, leading to respiratory and circulatory collapse. These are agents intended to kill or severely disable the target. The Japanese cult Aum Shinrikyo, after several failures, succeeded in a 1995 terror attack in Tokyo using sarin (the attack occurred in the subway, a confined airspace where the concentration of the gas would be manageable).

This sounds frightening, and it is, but the fright should be tempered. As I said in the beginning, gas is hard to do well. Theoretically, a quart/liter of nerve agent contains approximately 1 million lethal doses; in practice it requires roughly a ton/tonne to kill 10,000 people outdoors. Aum Shinrikyo tried several times, with far better-educated technical specialists than most terror organizations have at their disposal, before they managed to succeed in one attack, which had far fewer casualties than an unsophisticated bomb.

You should not ignore a threat though, historically, most gas "attacks" have been hoaxes. The same is true of biological threats.

If gas is hard to do well, biological attacks are even harder. Biological attacks take longer to develop, due to the incubation time in the victim. Biological agents may or may not be communicable (that is, they may require a means of dispersal beyond people infecting each other). Infectious agents must also be prepared to the point of infectiousness (without infecting the preparers, of course) and then kept in that condition until delivery. Living organisms are frankly much more delicate than dumb chemicals (and nerve agents especially are often set up as two inert chemicals that react and give off their deadly gas only when combined—one of the problems with delivery). Bacterial or viral agents have a far narrower range of ambient conditions within which they must be maintained, or they die. This is not a home-brew-in-the-garage operation that just anyone can do.

Biological warfare, in fact, is just that—warfare—and it is most likely to be conducted by an existing political entity (such as a rogue state; the United States Department of State maintains a list of such rogue states, available in its Overview of State-Sponsored Terrorism). Biological attacks by nonstate-sponsored groups is unlikely. Despite the anthrax-tainted letters sent in the aftermath of September 11, 2001, this is an extremely low-probability event; any preparation you undertake for other forms of attack should serve to ensure your network's survival and continued functioning in the event of a biological attack.

At the time of this writing, the profile of the person most likely to have committed the anthrax attacks is what might be called a *biowar hacker*. This person knows the material intimately from having been a part of the system but believes we are unprepared for biological warfare. He or she has sent the letters (taped shut to prevent accidental leakage, though no one apparently realized the problem of multimicron-sized pores in the envelope through which spores could pass) to show us how vulnerable we really are. It was intended to be a wake-up call to how easily a biological agent could be distributed (like the hacker who disrupts your network to show you which opening you have left). A total of fewer than one dozen letters appears to have been sent, probably all by one individual. The U.S. Postal Inspection Service has also responded to more than 15,000 hoaxes in the few months since the letters first appeared.

The R in CBR is for radiological attacks. This does not mean a nuclear explosion, but rather a deliberate use of radioactive material to contaminate an area. Unlike a nuclear weapon, where explosives surround the radioactive material and implode it to create the necessary conditions for the nuclear explosion, a radiological weapon has nuclear material surrounding conventional explosives. When the latter are detonated, the nuclear material is spread over a large area; some may well be small enough to be wind-borne and contaminate an extended region. The falling dust contamination has led to the nickname *dirty bomb* for such a device.

Like biological weapons, radiological attacks require access to controlled materials (germs, like nuclear materials, are supposed to be highly secured), but sadly, we know that those are not well-enough controlled in all areas of the world. Unlike biological weapons, it requires no great expertise to make a radiological bomb; any competent bomber could build it if the necessary material came to hand. In terms of your network's survival, you have two issues: First, there is the bombing itself; second, you would have to deal with the contamination.

If your network is architected to survive a conventional bomb, a dirty bomb explosion will be no different. As for the contamination, like a biological weapon, this is an issue that develops over some period of time, enabling you to perform a graceful transfer of functions to your alternate facilities.

Physical attacks, then, require the same sorts of redundancy and survivability preparedness as natural disasters. You may experience the loss of access or the loss of any operating ability at an entire location. To see how this worked and didn't work on September 11, 2001, we can examine a few of the many cases from the loss of the World Trade Center.

World Trade Center Examples

Timelines available from media sources vary a little bit, but they are within a few minutes of each other. For convenience of aligning times, they are listed here in 24-hour clock format (EDT).

0845	North Tower (1 WTC or Tower One) struck
0903	South Tower (2 WTC or Tower Two) struck (lower, angled)
1005	South Tower collapses
1028	North Tower collapses

The North Tower, though struck first, lasted longer—approximately 105 minutes. The South Tower lasted 62 minutes. Both towers were 110 stories high and not as fully populated as they would have been even one hour later. In the aggregate, approximately 18,000 people were evacuated from the two towers in the time available, via stairwells (very roughly, 100 people per minute for each building), while rescuers used those same stairwells in the opposite direction to enter and go up looking for people who needed help. Hundreds of firefighters, police officers, and rescue crew perished in carrying out that duty.

The possibility of a terror attack on the complex was not unexpected; a security review by a consultant for the Port Authority of New York and New Jersey (the buildings' owner) in 1985 defined four levels of terrorism and gave an example of each type applied to the towers:

- Predictable attacks: bomb threats
- Probable attacks: bombing attempts or computer crime
- Possible attacks: hostage taking
- Catastrophic attacks: aerial bombing or chemical agents in the water or air conditioning

The same 1985 report warned that the most vulnerable part of the towers was their underground parking garages; that was only rectified after the 1993 bombing, which could have been far worse (the bombers were unable to park in their preferred location, which would have damaged the building's supports much more seriously).

The report's author also warned in a 2000 review of the Port Authority's procedures about the possibility of a catastrophic airplane crash into one of the buildings (I have seen nothing to indicate that even he imagined such an attack on both towers) or someone blowing up the subway tubes under the buildings, followed by flooding from the Hudson River (a membrane held the water back from the foundations).

Is there a similar warning report about your facilities, gathering dust until after the fact?

Successes

The problem is that we cannot prevent a terror attack without becoming something other than a free and open society. Given that, it behooves us to be prepared to respond and keep our lives and businesses going as best we can in the immediate aftermath. Here are some examples of planning that worked; we'll follow with some examples of the struggles caused by the lack of adequate plans.

NYBOT

The New York Board of Trade is an exchange where contracts for future delivery of agricultural commodities, such as coffee, cocoa, and cotton, and financial instruments, such as U.S. Treasury Notes, are traded. It was housed in 4 World Trade Center (see Figure 7.2).

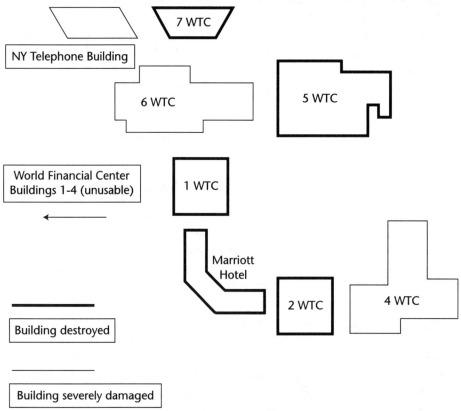

Figure 7.2 WTC and Telephone Building.

Its chief executive saw flames explode from the side of the North Tower and ordered an immediate evacuation of the NYBOT. He attributes his decision to innate caution; whatever its source, the trading pits were rubble within a few hours. For those not familiar with them, futures exchanges trade face-to-face in a pit by open outcry. Orders are telephoned in and confirmations telephoned back out to member firm offices where the orders have been placed by customers. Automated displays track the minute gyrations of prices on multiple active contracts for each subject of trade (Dec 01 Cocoa, for instance). Prices can move far faster than all but the most violent stock market activity; the communications and facilities for such an exchange are not easily created from thin air.

The NYBOT didn't have to. For the prior five years, it maintained a set of backup facilities about four miles away, across the river from Manhattan in Queens. It cost $300,000 per year, an expense not considered necessarily a wise use of the Board's financial resources by some of its 1,200 members. Needless to say, their opinion seems to have changed. NYBOT leased the backup facilities from Comdisco, Inc., a disaster recovery firm. The alternate facilities were not a full replica of the original; instead of 13 trading pits, they made do with 2, and therefore the trading hours of a given market were reduced so that the pits could be shared.

COMDISCO, SUNGARD, AND SEPTEMBER 11

Comdisco, Inc. has been in the disaster recovery business for several years. An IT services firm, disaster recovery is only one of the services it has offered on a global basis. Its centers are often located in unassuming buildings in suburbs rather than in cities; it offers a complete business continuity set of communications and workspaces, equipped with PCs as necessary.

In July 2001, in response to weak market conditions, Comdisco filed a Chapter 11 bankruptcy petition in the federal court in Chicago. The filing did not affect Comdisco's foreign business operations. A competitor, SunGard, petitioned to buy Comdisco's recovery services; by mid-November, the sale was completed despite U.S. Department of Justice objections concerning market concentration.

Between SunGard and Comdisco, more than 70 firms made use of planned backup facilities (in a total of 19 different locations) as a result of the events of September 11. These numbers exceeded their use by firms after Hurricane Floyd in 1999, which damaged a far larger geographic area. Many firms were back in operation by Friday, September 14; one financial services firm was operating well enough to trade $6 billion in bonds on Wednesday, September 12. Most were able to locate and set up more long-term new locations by the end of September.

Three weeks of operation, even on a limited basis, is better than three weeks of lost business, especially in an information-intensive enterprise where you really have few tangible assets. The Gartner Group has reported that 40 percent of all businesses that experience a disaster such as the loss of a major operating location like the World Trade Center go out of business within five years.

Communications did not come up seamlessly when the alternate location began to come online. NYBOT's ticker stream from Global Crossing had to be rerouted because the latter's data center, which was its normal source, had been destroyed. Connections had to be rebuilt to member trading firms, some of whom were also operating from temporary facilities. By Saturday, they tested everything for a Monday resumption of trading. Telephones seemed to work fine until they tried picking up every phone at the same time, but that failed due to a local equipment glitch that was found and repaired in relatively short order.

The Wall Street Journal

The Wall Street Journal published the story of its own recovery on October 8 (modestly enough, after spending a considerable amount of copy on the problems and recoveries of other businesses). But its story is useful in that a newspaper is a quintessential information-based business. The newspaper had learned from its own prior experience—not with the 1993 WTC bombing, but with New York's 1990 power blackouts—that distributed operations could be its safety net.

During those blackouts, it had to rely on emergency generators (probably diesel, though it did not specify). The newspaper had a fire in the generator room, which put its entire emergency power supply at risk of failure. Without that power, and with all operations centralized in its Manhattan offices due to manual paste-up of the paper galleys, it could not produce its product.

A day late is too late in the newspaper business. *The Wall Street Journal* moved to electronic pagination, at a location outside New York, and decided that it would never be limited to only one location where it could put out the paper. That decision, and the use it made of those redundant facilities, meant that it was able to deliver a product—a full newspaper, with eyewitness accounts of the tragedies and amplifying information on the political and economic situation—on time, the next day.

From their offices in a building across the street (offices that were functionally wrecked when the towers collapsed), senior managers could see the towers burning. The managing editor asked whether they could produce a paper if they evacuated; the answer was yes, if the pagination facility had its test newsroom up. A quick call verified that it was; personnel there had been watching the news and had already begun activating the features necessary to support main operations. The pagination facility is about 50 miles away; it has become a grind to work from there and live in New York, but the distance is a reasonable assurance of safety from simultaneous disasters. We'll address that concept again in Chapter 9, "Preparing for Disaster," and Chapter 10, "Returning from the Wilderness." With confirmation that the capability would be there, management ordered an evacuation of the current location, a move to the alternate location—and a six-column headline. (In the midst of a disaster, wouldn't you like the luxury of thinking about how best to deliver your product the next day instead of whether you can deliver at all?)

Not everyone made it to the alternate location; many worked from home or from the homes of colleagues who had high-speed Internet connections (mostly DSL). The Washington bureau also picked up some of New York's ordinary editing duties. Information was exchanged via email because long-distance lines were out (a problem of congestion and lost connectivity, depending on location). Email through in-house accounts worked because, by luck, the *Journal*'s data center was on the far side of the building and the servers ran for three days on their emergency generators, which kicked right in when power was lost. By the end of the three days, systems had been rebuilt to work from the alternate location.

In another stroke of luck, due to some testing the pagination system was already operating on the backup servers, which saved several hours in switchover time. Something that was definitely not luck was that the alternate location had been made the technological center of the *Journal*'s operations, a center that manages the infrastructure required to allow it to print region-tailored editions from 17 locations, as well as the electronic edition. The statistical section of the paper has been routinely composed there every night, for the specific purpose of keeping the circuits used and a cadre of people in place.

In short, *The Wall Street Journal* had an alternate location, with the technical capacity to support operations and with connectivity and a staff deliberately located there just in case. With that in place, management did not have to think about lost business or other costs; they could be focused on the needs of the moment—to get their people out. They had creative people who found ways to get the job done, not necessarily from an office. They had a little luck—but luck can be defined as the intersection of preparation and opportunity; without preparation, the luck cannot be used to advantage or possibly even used at all.

Lehman Brothers

Lehman Brothers is a major financial services firm, with 45 branch offices outside New York. Its headquarters were in 1, 2, and 3 World Financial Center (WFC; across the street to the west from the towers) and in Tower One. In the latter, the IT department, including the technology management team, was located in offices on floors 38-40. When the CTO saw the roofs of surrounding buildings on fire and debris falling from above on the tower, plus water shooting out of the elevator banks in the lobby, he ordered an evacuation. Only one of their 625 employees in Tower One failed to make it out.

During the evacuation and the subsequent chaos, the only communication available was email via BlackBerry; with that, the CTO triggered Lehman's disaster recovery plan and alerted the CIO, who was working from London on September 11.

The financial markets had opened, and were later closed, but trades had been made. Lehman's treasury department moved to the backup recovery location and performed its cash management functions the same day. The following day (Wednesday), the company was online trading fixed-income securities (which do not trade via a centralized exchange like stocks or futures). They also had 400 traders online when the NYSE reopened the following Monday.

In addition to the one lost individual and the trauma inflicted on so many others, Lehman lost a total of 5,000 desktops between the WTC and the unusable buildings of the WFC, its entire New York data center, and all the associated networking equipment. While it may have lost a few other things, the company lost no information. Its disaster recovery plan ensured complete redundancy of that, its corporate life. In short, the plan assured Lehman's continuity.

The backup facility is across the Hudson River, in New Jersey. A fiber link connected Tower One with the offices in the WFC across the street and continued to the backup site. The wide-area links were redundant, and the backup site had a data center identical to the one in Tower One. All branch offices connected to both data centers, and the centers replicated to each other automatically.

Lehman's disaster plan allowed for the complete loss of either of the two data centers. When the one in New York was lost, part of the recovery plan was to increase bandwidth to the other to handle all the branch office traffic instead of only a portion. Another part was the establishment of a new redundancy, this time between New Jersey and the data center in London.

Lehman worked out a deal with a hotel in Manhattan to set up office there, at least temporarily. A VPN was set up from the New Jersey data center to the hotel. Unfortunately, one thing not anticipated was the insufficient supply of connectivity at the hotel. Lehman filled all 650 rooms of the hotel with its bankers and their staffs; however, being a hotel, the phone system was designed to handle about only 75 outgoing calls at any one time and had limited data capacity. It normally does not need more than that. Verizon, the local telephone provider, managed to increase capacity to 300 voice lines within a few days, with further increases to 500 not long thereafter. Verizon also assisted in providing increased data connectivity.

All three success stories here share some fundamental characteristics:

- They all had a disaster recovery plan in place, with facilities already available to handle the complete loss of the ability to operate in the primary location.
- Those plans were activated promptly by senior management, who also promptly got their people out of the buildings.
  - Lost equipment can be replaced far faster than someone who knows how your business and your network operate.
- Not everything went according to plan, so adaptations had to be made.
  - Voice communications turned out to be less than needed in initial configurations at two of the three backup locations.
  - The third was lucky enough to have a three-day window to reconfigure email services as a result of their primary location's server placement within the building.
  - People wound up operating from unexpected places (homes, hotels) using electronic connectivity via Internet, email, and even BlackBerry.

Even in a disaster of lesser magnitude, such as major damage to a building (yours, unfortunately) and no major communications outages, would your firm do this well?

Lost Access

We have seen that there were problems of insufficient phone connections at the planned backup sites. There has also been a great deal of information in the news concerning how much telephone and data connectivity was lost.

Bear in mind that, in North America, there is generally one major local provider, and it is a successor to the local company formed in the breakup of AT&T under the consent decree that ended the federal government's monopoly litigation. In the New York area, that provider is Verizon, which was formed by the merger of GTE with Bell Atlantic, which had already merged with NYNEX. Competitive local providers usually do not create entirely new local loops; they connect their switching equipment to the existing voice network at the incumbent's Central Office (CO).

Therefore, local voice networks still depend on the network developed in the monopoly era. As Shawn Young and Dennis K. Berman described the infrastructure in Manhattan in *The Wall Street Journal*, "The New York City telecom system is itself a relic of historical quirks, with different systems built on top of one another, forming a kind of technology sediment."

The major switching center for Lower Manhattan was in the New York Telephone Building immediately west of 7 WTC, at 140 West Street. Switches inside the 32-story building carried roughly 30 percent of all the area's traffic (voice and data), a load comparable to the entire voice and data traffic of a city the size of Cincinnati; the connections totaled approximately 4.5 million data circuits and 300,000 voice lines. This is Verizon's description to the Federal Communications Commission of their damage:

The damage to 140 West St. was extensive. The building exterior was penetrated on every floor on which we have equipment. Steel girders pierced the building's façade, cable entrances were crushed, and a water main break flooded sub-basements. Modules from a 5ESS switch [a very large voice switch] fell into the debris, power generators were under water, the building was coated inside and out with soot and ash, and the main cable vault under the building was under four feet of water. The air pressure system failed allowing water to move through the cables into duct banks and manholes under the rubble.

As a result of the damage, some 300,000 dial tone lines and some 3.6 million DS-0 equivalent [64kbps] data circuits were out of service, affecting approximately 14,000 business and 20,000 residence customers.

When 7 WTC collapsed, steel girders punched through the third floor exterior walls of 140 West Street and crashed through the building to the second basement level. A gaping hole on the seventh floor exposed to the soot and ash the communication switches dedicated to a major banking/financial market maker.

By Friday, September 14, Verizon had rerouted the equivalent of 2.1 million voice and data lines around Manhattan. As noted, a very high number of manholes were buried by rubble, and the conduits beneath them were inaccessible and so could not be repaired and/or pumped out to be repaired. Verizon increased wireless service by bringing in portable cell towers, and it ran fiber cables in conduit above ground, in hastily dug trenches, and through windows to restore connectivity to the New York Stock Exchange so that it could open by Monday, September 17. Verizon employees hand-carried servers down 23 flights of stairs—wearing respirators because of the amount of dust, smoke, and hazardous material still in the air—to provide the price quotation system.

Virtually all of the fiber and copper lines that ran beneath the streets in Lower Manhattan have had to be replaced; aside from the damage caused by debris falls, substantial flooding occurred due to the firefighting required for months—literally—before the fires resulting from the attack were declared out. As of October 19, for instance, roughly 16,000 lines connecting the Chinatown area to the rest of the world were still out because they ran in conduits that passed directly beneath the WTC, and they were still both buried and waterlogged. By November 30—11 weeks after the attacks—Verizon estimated that approximately 8,000 circuits were still not functioning.

Verizon estimates that it will spend at least $1 billion to repair and rebuild the local communications network. But it was not the only one that lost circuits. Metromedia Fiber Network Services, Inc. (MFN), for instance, filed a report with the Federal Communications Commission indicating that it had lost the following (destroyed; not repairable):

- 2 optical cables containing a combined 312 fibers; these were used to connect MFN's customers within the WTC and elsewhere in Lower Manhattan to MFN's backbone.
- 144 optical fibers that connected the backbone to 50 Broad Street, a major hub for MFN; these were destroyed during initial rescue and recovery operations.
- 864 optical fibers connecting MFN's backbone to Verizon's CO at 140 West Street.

That is a total of 1,320 optical fibers that would require replacement ... once the company could get physical access to the areas where they needed to be run, and once those areas were safe to use (not flooded, for example), and the appropriate permissions had been obtained (for new routings), and so on.

This is enough information concerning the magnitude of the outages for you to realize that, even if your firm's physical property is not in the immediate disaster zone, your connectivity to the outside world can be disrupted, and for a substantial amount of time. Your disaster planning must take that into consideration.

Less Than Successes

Finally, as painful as it may be to show a few organizations that made some mistakes in their preparation, we may learn some valuable lessons from those mistakes. I do not necessarily characterize these as failures; in two of the three cases I cite, personnel performed beyond expectation to make do with what turned out to be a resource set far less than what they needed. In the third case, that information is not available, for reasons that will be apparent. As a result, I am characterizing these as "less than successes" because, with better planning and resources, those same people could have done far more. These situations really come down to two issues: lack of planning and lack of capability development even though a need had been considered.

The Local Loop

Verizon, as noted before, was the successor company (through many steps) to the original New York Telephone Company, which built the art deco skyscraper at 140 West Street. And compared to what happened in Chicago in 1988, when a relatively limited fire inside a switching station closed O'Hare Airport (on Mother's Day) and left 38,000 local users without service for as long as a month, Verizon did enormously better. This facility, I must emphasize, was no different in concept than a CO in any other population center; the magnitude of the problem arose from the sheer concentration of users—data and voice—in the capital of capitalism, Lower Manhattan.

It is no different elsewhere. If one building in a certain western U.S. state were destroyed, for instance, not only would that city lose all local service, the entire state would be cut off from the long-distance network. Why are there still such single points of failure in the telephone network? Competitive Local Exchange Carriers (CLECs) have been able to offer service while providing connectivity only from the CO on into the network because the Incumbent LEC (ILEC), almost always a Bell System inheritor company, has been required to allow them to use the existing local loop. In order to operate as a telephone company, the CLECs have not needed to invest in installing a separate local loop of their own; prudently (in capital terms, anyway), they have not. When a consulting group examined the issue for Amtrak, it found that less than 10 percent of the CLECs had truly separate networks from the ILEC. Even had they chosen to invest the capital, they may well have run their cables and fibers through the same conduits underground; we'll come back to this right-of-way problem in Chapter 8, "Unnatural Disasters (Unintentional)."

On a network scale, the local loop is currently a single point of failure unless you take very great pains to ensure redundancy. As the CTO of a firm displaced by the disaster commented, "It's just something that wasn't worth the cost before the unthinkable happened." Lots of firms are thinking about it now and finding it extremely difficult to be sure that they have Physical Layer redundancy. Further, even if you contract with two separate LECs whose cables enter the building on opposite sides, and even if you have a circuit diagram that shows separate connectivity paths all the way out into the carrier backbone (usually depicted as a cloud on network diagrams), can you be sure it stays that way?

A common practice in the communications industry is grooming, which (among other things) tends to consolidate traffic on fewer physical links in order to make better use of that capital asset. If you buy service from ABC Telco and from DEF Telco, they may both be purchasing connectivity from GHI Networks. If GHI grooms the circuits and combines the incoming traffic paths onto one fiber exiting its CO, for instance, a simple fiber cut may take out both your (redundant) telephone connections beyond the CO where the grooming occurred. In a case like the attack on the WTC, you could have had service from both Verizon and AT&T and still lost connectivity because both lost switching centers (AT&T's was in one of the towers). We will see a similar situation in Chapter 8 when a fire in a train tunnel melted the fibers of multiple carriers.

As Sean Donelan, a well-respected voice in the North American Network Operators' Group (NANOG), www.merit.edu/mail.archives/nanog/, wrote on December 29, 2001:

> *We don't wait for buildings to burn down before checking the electrical wiring. We don't take the electrician's word the wiring is correct. A building inspector checks the wiring before the walls are covered up. Its not perfect, and the inspectors can miss alot. But it's proven to be more effective than trusting the builder will do the right thing.*
>
> *Where are the building inspectors for the Internet?*
>
> *As the insurance industry figured out, SLAs alone don't improve things. You have to inspect and verify. How can I inspect and verify a carrier did what they promised? If I pay for a diverse circuit, can I check it is in fact diverse?*
>
> *Is it a matter of money? The Chicago Board of Trade had a multi-day network failure. The New York Stock Exchange had a multi-day network failure. The NASDAQ had multiple shorter network failures. Essentially all of them traced back to carrier problems.*
>
> *You could hire me to inspect your network, but that doesn't scale.*

Who is the building inspector for your local loop?

If you want to be sure you can use your telephones and data circuits in the event of a terrorist attack, you will have to do some auditing, and don't assume everything is good beyond the CO just because your contract says something about diverse circuits. Check them. And you will have to do this more than once every year or two, when somebody needs a project to occupy some time (or as payback for some faux pas). We'll return to this topic in Chapters 9 and 10.

New York City OEM

The City of New York Office of Emergency Management had a state-of-the-art emergency command post, often called a bunker, in 7 WTC. When that building was evacuated due to damage from the collapse of its neighbors across Vesey Street, the city lost its one facility intended to handle just such an event. New York's Deputy Mayor of Operations, in speaking to a conference sponsored by an alliance of several national associations of state and local government officials, admitted that the city had to scramble to create a makeshift command center during the crisis (and both City Hall and the Municipal Building had been evacuated as well). City officials had not thought they needed a redundant command post because the one inside 7 WTC was built to withstand earthquakes, hurricanes, and any other natural disaster New York might have experienced, as well as a bomb.

In fact, there is significant evidence that the presence of the command post may have been part of the reason the building was lost. As of November 29, structural engineers knew the building came down because of the uncontrollable fires that raged within it (the Fire Department had been forced to abandon any attempts to save it one hour after the North Tower, nearest it, came down). But 6 WTC, which actually stood between Tower One and 7 WTC, while severely damaged, did not burn to that degree and did not collapse (see Figure 7.2). Likewise, much of 4 WTC and 5 WTC still stood, even though they, too, were plagued by fires.

The questions are where did the uncontrollable fires in 7 WTC come from, and were they the only reason for the collapse? Or did falling beams and other debris damage the supports, and the fire simply finished the job?

The answer to where the fires came from is probably (though not certainly, as this is written) the diesel fuel supply for the command bunker's generators. The diesel fuel tanks were surrounded by fireproofed enclosures. Whether those were penetrated by falling debris is a matter for forensic analysis. A small tank on the seventh floor of the 47-story building served the generators' immediate needs; it was resupplied by a larger tank on the second floor that held 6,000 gallons. That, in turn, was supported by four tanks holding a total of 36,000 gallons of diesel fuel; these were just below ground level at the southwest corner of the building.

The American Society of Civil Engineers and the Federal Emergency Management Agency have a team evaluating why the building collapsed (some steel members in the debris pile were partly evaporated by the high temperatures). Their concern is that many facilities nationwide have diesel fuel supplies for emergency power; if that is the only explanation for what happened to 7 WTC, we may have to rethink how we prepare continuous-operation facilities for power outages. You may need to rethink yours.

The good news in all of this is that, despite the lack of planning for a backup command center, the planning the city did for the Year 2000 problem stood it in good stead. It had forced the city to go through the agency's critical and operational issues, and those lessons had not been lost. It cobbled together, with donated and borrowed equipment, a network that supported the disaster recovery until a more sustainable one could be created.

The U.S. Secret Service

The U. S. Secret Service does more, much more, than protect the President and other senior officials in the Executive Branch and visiting foreign dignitaries. A part of the Department of the Treasury, it is the lead law enforcement service for counterfeiting, and it also has responsibilities for investigating credit/debit card fraud, federal-interest computer fraud, identification document fraud, and civil as well as criminal investigations related to federally insured financial institutions.

A USSS office in the Alfred P. Murrah Federal Building in Oklahoma City was lost when a terrorist bomb destroyed the building in 1995. The loss of paper files and nonredundant soft files led the Service to develop plans for better backup of all information. Unfortunately, despite having had experience with data loss, it received no funding with which to implement the plan for its New York office in the WTC.

We can say in hindsight that the USSS should have reprogrammed money within its budget to achieve the goal anyway; that is not always an option for a government agency. An unknown number of federal investigations will have to be done over, if the evidence can be reacquired (perhaps the only lead information was on a paper file), and when resources can be made available.

These six examples and the problem of completely lost access remind us that we must do our planning based on total loss of our facility's capability (whether the building itself goes out or we simply lose all communication with it). They also serve to point out that information continuity requires preplanned and already extant facilities, preferably with at least some people already in place (for other reasons, perhaps) who can carry things until help can arrive. Planning alone is not enough; the funds must be spent both in acquisition and in continuing operating funding, or when you need it, the capability will simply not be there.

Cyber-Attacks

Of course, a terrorist attack need not be so crude as a bomb or anything physically dangerous to the perpetrator. It could be a form of network attack. Terrorist organizations have often relied on criminal activities to pay their ordinary expenses. In that regard, we could expect the cyber equivalent of kidnappings and extortion to be used against firms believed to have deep enough pockets and an unwillingness to have their security flaws exposed.

Cyber-Kidnapping

What would a cyber-kidnapping be like? Consider how a human kidnapping works: The criminal removes from safety and places in danger not the person with the checkbook, but rather someone the person with the checkbook cares about more than he or she cares about money. That is a simplification, of course, but it captures the essence of what occurs—someone of greater value than money is placed in a position from which he or she can be redeemed only with money (and a great deal of it).

A cyber-kidnapping then would entail the removal of files, such as critical information on a research project or new product. Such events, involving stealing source code, have already occurred. Your sensitive files might well be kept on a special portion of your network; if local administration is negligent or lazy or even ignorant of corporate procedures (perhaps due to downsizing, the person who inherited the position is underqualified and overwhelmed), it is conceivable that you could discover a proper backup was never made or not made very recently (and much work may well be lost). The files would then be ransomed.

It is easy to assume that the backups would render this moot—the kidnapper could not take the only copy of the information that you have. Perhaps, and perhaps not. If the kidnapper removed the files and also managed to get at the backups (through insiders, perhaps—an appalling proportion of human kidnappings involve the assistance of an insider), what recourse do you have?

It is unlikely, to be sure, but do not assume that it is impossible.

Extortion

Threatening to damage your system (with some token damage, to demonstrate both capability and willingness) is more likely. Threatening to remove those critical files is more likely than their actual removal. There is also the possibility of a combination of the characteristics of the two, in the form of removal (probably without being able to affect your backups), with the threat to sell the information to your competitors or expose the data unless you ransom the files and/or pay the extortion.

As discussed earlier, once you pay the Danegeld, you never get rid of the Dane. You have no real assurance in this situation that the criminal hasn't kept a copy of the file (kept the negatives, as it were, of the incriminating photos) and will continue to hit you up for more payoff, until the information has lost its value. Likewise, in this instance, as in human kidnappings, just because you pay a criminal does not mean you will get what you paid for (the return of your precious information). After all, what can you do to them? Sue?

Easier Targets

Historically, critical infrastructure systems have not only been less exposed to outside networks, such as the Internet, but they have also been built from less common software. This may be software written in old languages (finding programmers who could work in COBOL was a challenge in the run up to January 1, 2000, simply because it is not used enough to be taught in colleges; easier and more versatile languages have largely replaced it) as well as proprietary software that is of no real value to anyone else and so has remained essentially unknown. Likewise, network structures for these systems have remained obscure. Because of this, cyber-terrorism against infrastructure, such as the electrical grid or the underlying telephone network, has been a nontrivial exercise.

As a part of rationalization as well as modernization, older, proprietary systems are gradually being replaced by systems based on known operating systems (such as UNIX or Windows NT/2000); C or one of its derivatives is often the programming language of choice. And information is usually stored using commercially developed products from major software companies; those products must be reasonably well documented, and their characteristics (and weaknesses) are far more widely known. This makes a cyber-attack on infrastructure far more feasible.

And it may well have been done, once before, by hackers for fun.

Richard Power, in his book *Tangled Web: Tales of Digital Crime from the Shadows of Cyberspace*, offers interesting evidence that the crash of AT&T's

long-distance network on Martin Luther King Day in 1990 was not an ordinary software glitch. The network, he strongly suggests, was, in fact, deliberately taken down by a hacker. Those capabilities are available to anyone who studies the right material and has access (whether through human engineering, extortion of the appropriate individual, or simple purchase of the information from a disgruntled insider) to network structure information.

Terrorist organizations with a hatred for Western culture and deep pockets from their own financing or state sponsorship certainly have the will to do such a thing if they think of it. With patience, given the shortage of IT personnel, they could place one of their own in position to acquire the information/plant a back door/discover the network, and complete the information set needed for the attack.

Combined Attacks

Imagine a physical attack in a major population center. Add a cyber-attack that, if nothing else, clogs the Internet due to a worm or DDoS attack. Many businesses in this major (trophy) property have a converged network, using Voice over IP; perhaps the cyber-attack includes malformed UDP packets clogging the port used for VoIP.

What happens to the communications needed to respond to the physical attack? What if the cyber-attack took down major portions of the telephone network (as in the MLK Day crash), and the government/emergency responders had to fight the congestion of everyone trying to get in touch with others?

After all, that's exactly what I did when I was in a building bombed by terrorists. I secured my classified material; with two others in the office, we made the rounds to ensure that all classified material in the office was secured, and then I ran across the street to call my husband and tell him I was fine. This is what everyone tries to do in such a situation—reassure each other of the safety of those we care about. As AT&T stated in its report to the FCC:

> *AT&T experienced multiple service disruptions in the New York City and Washington, D.C. areas as a result of the terrorist attacks at the World Trade Center (WTC) and the United States Pentagon. Extensive infrastructure damage impacted four local New York City switches and triggered mass call events. Nearly a dozen 4ESS switches experienced periodic intermittent overload conditions until network management controls were implemented to help alleviate traffic congestion.*

The report estimated that AT&T experienced a total of 111 million blocked calls, and the outage affected 37 million customers (a number AT&T felt was probably high, but it had been forced to estimate it using an algorithm).

Thinking about these what-ifs may sound like something the government ought to be worried about, and it is. But how does that help your network survive? It doesn't; a continuity plan on the order of that which saved Lehman Brothers and *The Wall Street Journal* would because the replication was already done. The data was already where it could be accessed; the alternate communications were already in place.

During the chaos that followed the attacks in New York and the building collapses, a number of people who made corporate decisions were able to disseminate those decisions only via their BlackBerry devices. Made by Research in Motion (RIM), the devices operate on a different portion of the spectrum (paging) than that clogged with people trying to make cellular calls. Of course, because this has become known, more companies may employ the devices, and (conceivably) that part of the spectrum could become clogged, too. Further, one of RIM's two nationwide carriers, Motient, Inc., has filed for Chapter 11 bankruptcy.

Several firms whose survival stories have been documented noted that they exchanged emails to get word out; this was possible because the Internet and communications backbones were not deliberately attacked. But the point is that, when the first choice of connectivity didn't work, another was tried, and another, and so on. Different forms of communication, using different connectivity, made the information exchange vital to critical business decisions possible.

That is sort of what the network is there for.

You never know when you will need that, so you must have it available already. Several years back, an Irish terrorist taunted the Scotland Yard Inspector with a football (soccer) analogy. The police, you see, were like the goalkeeper—they had to be right every time. The terrorist had to be right only once.

Once is all it takes; just ask the firms in the WTC that did not have continuity plans, or that were so physically concentrated on the upper floors that their most precious human capital did not survive.

CHAPTER 8

Unnatural Disasters (Unintentional)

Shallow men believe in luck. Strong men believe in cause and effect.
Ralph Waldo Emerson

Terrorism intends to disrupt, to sow confusion and doubt. It occurs, really, not very often. Bad luck, unfortunately (so to speak), is far more common.

Luck, of course, has been defined (at least as far back as the Roman statesman and philosopher Seneca) as the intersection of preparation and opportunity. That is true of bad luck as well as good. If you have not prepared the ground, the seed of an opportunity for trouble will not sprout. From that perspective, there are three areas to consider as you try to make your network more survivable: unfortunate opportunities, unfortunate planning, and unfortunate implementation.

Unfortunate Opportunities

Unfortunate opportunities are the things that most people call bad luck. One line of investigation after an airplane crash is always the possibility of a bird strike (though, when you think about it, it was more likely the airplane that struck the bird rather than the other way around). These really do happen, and the damage is a function of the size of the bird as well as where it impacted the plane. If a large bird, such as one of the waterfowl species, strikes the engine of a jet, that engine will be destroyed; the destruction generally results not so much from the actual impact of the bird as from the initial damage spewing little pieces of former bird and former intake into excruciatingly fine-tolerance turbines spinning at many thousand rpm.

The small misfortune cascades into a much more serious problem.

This is certainly bad luck for the bird and for the airplane, which must now be brought in with asymmetric thrust (if it wasn't a single engine, of course) and probably damage to the hydraulic and electronic controls. Because this is a chancy proposition for those aboard the aircraft as well as those on the ground in the proximity of its semi-controlled (if that) landing, aviation authorities go to great pains to keep birds away from airports, where the airplanes are most likely to be at an altitude to be struck.

We try to prevent the unfortunate opportunity from arising.

Unfortunate opportunities (henceforth just "bad luck") are those events that affect you and over which you have no control (like an aircraft striking a bird—nothing the pilot could do could have prevented the impact). Perhaps a backhoe sliced a fiber on your provider's backbone; your connectivity may be affected, yet you could have done nothing to prevent the incident from occurring in the first place or impacting your network in the second (as long as you have made the business decision to connect via this provider). Likewise, the downtown flooding incidents discussed in Chapter 5, "'CQD ... MGY'," were bad luck to any firms that lost connectivity because the conduits carrying their physical cables were unusable.

Reportable Outages: They're Everywhere

LECs in the United States are required to report service-affecting outages to the Federal Communications Commission. In 2001, a total of 200 outages were reported. For instance, picking a report at random, we see that Qwest had an outage on December 6 at 10:36 A.M., caused by an AC power failure. Duration of the outage was 1:55, and Qwest estimated 47,466 customers were affected, with an estimated 76,654 calls blocked (in telecom, a circuit setup is known as a call), just under 40,000 calls per hour.

Likewise, Sprint reported a fiber spur line cut on October 20 by boring equipment; being a spur, no backup route was available for traffic. The cut was repaired after a total outage of just under 25 hours; all interexchange, long-haul, and switched traffic was affected (a total of slightly fewer than 800,000 calls).

> **PREVENTING BIRDSTRIKE DAMAGE**
>
> A birdstrike on the cockpit can be devastating because the humans inside are often injured, and the cockpit is a mess of broken glass and animal and human blood, parts of the former bird, and papers, all blown around by a 300 mph/500 kph wind. It can be extremely difficult to return the airplane safely to the ground, especially if some of the debris ends up jammed into the joint where the control yoke inserts and thereby restricts its movement. Because of this, a great deal of testing is done on canopies and windshields to be sure they can withstand the energy transfer of a sizable bird at aircraft impact speeds.
>
> Boeing Aircraft used a cannon to fire a chicken carcass at a windscreen to test its durability, and others who learned of the test thought it a good simulation. A British firm tried it on their new, improved windscreen product; to their dismay, the chicken shattered the glass. After repeated testing to be sure it was not a fluke, the company wrote to Boeing and provided full documentation of its test procedures, then asked what could account for its wildly different results.
>
> After studying their information, Boeing engineers replied, "Thaw the chicken."
>
> It's a funny story, but it illustrates an important concept: Be careful that your simulation of a real event accurately reflects the important characteristics of the real event. Frozen birds do not normally fly around aircraft. Core data routers with carrier-grade redundancy and reliability do not just break. If you want one to break as part of your simulation, give it a plausible reason to break.

Verizon reported an outage affecting over 35,000 customers and causing well over 97,000 blocked calls when a technician, in the process of performing some maintenance to repair Problem A, caused the more serious Problem B. In the subsequent investigation, Verizon reported that it discovered that the local topology had not been properly designed to standards. In the "Best Practices" section of its report, it quoted the guidance it had violated: "Link diversification validation should be performed at a minimum of twice a year, at least one of those validations shall include a physical validation of equipment compared to the recorded documentation of diversity." Validation is inconvenient and no fun for the validators; handling a preventable outage is less fun.

Interexchange Carriers (IXCs) do not have quite the same reporting requirements, so a browse of historical threads at network operator mailing lists (such as NANOG) will reveal many instances of connectivity losses that you will not find reported to the U.S. government. There is a preponderance of fiber cuts, as these problems tend to show up quickly. There are also a number of reports of software-caused problems; as stated before, not all problems will be discovered when a carrier or service provider tests new hardware or software in the lab. Sometimes the problems simply do not show up until the system is operating under real-world loads and operational stresses.

There are also outages caused by someone misconfiguring routing equipment (usually a router, sometimes a switch). The real flaw here often traces back to poor local design, which we will address shortly. And you will find outages that trace back to poor infrastructure design, as well.

Most bad luck, however, comes from outside events. In a six-week period from July 22 to September 1 of 2001, for instance, LECs reported 31 outages, with a further 7 data-only outages reported among NANOG members. Across the United States, there were 38 outages in 42 days. Nor are those numbers unusual; Table 8.1 lists the summary numbers for 2000 and 2001 (removing the reports directly attributable to the terrorist events of September 11, 2001; these are the ordinary level of outages).

There are no exempt carriers; all major and many new, competitive carriers reported outages in this two-year period. Taking only the inheritors of the Baby Bells and the three major long-distance carriers, we see the following numbers shown in Table 8.2.

In the long-distance arena, AT&T has the most extensive (and by far the oldest) network, and this is reflected in its many more outages. Of the ILECs, USWest inherited the physically largest but least populated area, giving it long distances available for fiber cuts to occur. The more urbanized ILECs, on the other hand, have denser networks and, in more urban environments, more diggers to try to keep away from its cables and fibers. They report roughly the same number of outages.

Table 8.1 Monthly Reportable Outages, 2000-2001

	JAN	FEB	MAR	APR	MAY	JUN	JUL	AUG	SEP	OCT	NOV	DEC
2000	14	13	19	9	20	23	21	25	20	20	19	21
2001*	14	17	8	20	10	24	20	22	20	18	4	14

* Does not include reports directly attributable to September 11, 2001, terrorist attacks

Table 8.2 Reportable Outages by Major Carriers, 2000-2001

PROVIDER	2000 OUTAGES	2001 OUTAGES*	TOTAL TWO-YEAR OUTAGES
AT&T	27	24	51
MCI/WorldCom**	8	10	18
Sprint	8	11	19
Verizon/BellAtlantic/GTE**	38	31	69
BellSouth	30	29	59
SBC/AmeriTech**	34	42	76
USWest/Qwest**	54	31	85

* Does not include reports directly attributable to September 11, 2001, terrorist attacks
**Merged companies

Route Diversity in Reality

Obtaining and maintaining route diversity are difficult. There are only so many rights-of-way, and accessing them can be difficult in ideal circumstances. When circumstances are less than ideal, the price for access may soar, if access is available at all.

You may expect diverse routes; it may be in your connectivity contracts. But because carriers lease bandwidth from each other in many areas in order to economically complete routes, there can be no guarantee that, at some point, physical diversity will not be lost. Even if you had diversity, route reconfigurations may occur for economic or physical layer reasons, causing your diverse carriers to suddenly (and probably unbeknown to you, and possibly to them) share one conduit or right-of-way.

You must factor into your network continuity planning the possibility that you will lose all network connectivity to any given site.

Fire

In addition to fiber cuts, other forms of bad luck can cause you to lose connectivity with a site. One is fire. You may have an operation occupying part of a building; as mentioned before, if there is a fire in another part of the building, you may sustain smoke and water damage. You will probably have to evacuate. When you can return depends on the amount of damage and its proximity to your location (assuming the building has not become unsafe or is no longer within building codes, city regulations, and so forth).

Of course, you may operate from a campus location, not a multitenant unit. Consider what happened to Los Alamos Laboratories in 1999: When a controlled burn became uncontrolled (and drastically so) in the federally managed forest nearby, the resulting blaze swept through portions of the lab facilities (local topography provides some natural firebreaks that saved major portions). The lab's network managers had to relocate their emergency communications system five times during the several-day event. Or recall the Tunnel Fire in Oakland, discussed in Chapter 5; in a drought, that lovely shrubbery between buildings on the campus can carry the fire from one building to another. It might be your bad luck to be a tenant in Building Six, and the fire started in Building Three.

> **ALL ROADS LEAD TO ROME**
>
> Installing new fiber routes, or even new copper cable in the local loop, takes longer than you might imagine. Odd and strange problems crop up. For instance, AT&T was unable to use its own right-of-way to lay new fiber for an expansion of its broadband network in the western United States. The reason? The now useless coaxial cable it had installed *in the same right-of-way* in the 1940s had been designated a historic landmark. AT&T was forced to acquire new rights-of-way and engineer around the "obstacle" of its own property.
>
> Governments have made acquiring such rights-of-way both more difficult and more expensive (when you do win access) for two reasons: One, it is a source of revenue that can be used for other purposes, and such purposes can always be found in a democracy or a kleptocracy or anything in between. The second reason is competition; when telecommunications was largely deregulated in 1996, everyone, it seemed, wanted to build out a new network, so there were many applicants for limited space (physically limited space under city streets and limited routes over which it was physically practical to run fiber between cities). From 1980 through 1996 (17 years), 52 million miles/83 million kilometers of fiber were laid in the United States. In the next four years, an additional 33 million miles/53 million kilometers were laid. Cities not only demand far greater care and repair of streets when they are cut, they even have begun restricting when streets may be cut for new installations—once every five years, for instance; if you don't build in that window, you lose the opportunity to build until the next window.
>
> *(continues)*

ALL ROADS LEAD TO ROME *(Continued)*

One rule of thumb is that the networking company will require one permit for every mile of route, whether on private land or public property. Right-of-way permits used to be 10 percent of the cost of a new network; that had risen to almost 20 percent by 2001. When Williams Communications finished its buildout of a 33,000-linear-mile/53-million-kilometer fiber network in December 2000, it had required approximately 31,000 right-of-way agreements and 10,000 government licenses and other permits from landowners. Aside from environmental obstacles (construction of certain segments may have to be timed to avoid disturbing the nesting patterns of birds, no matter what that does to the rest of the construction schedule), private landowners do not always understand what is going to be carried on the right-of-way. Teams have had to show rural landowners that the fiber cannot cause a spark, while some owners have objected to fiber carrying Internet pornography across their property.

When right-of-way has been acquired, it is often not yours alone. In fact, AT&T avoided installing its fiber along some state highways because the easements were already so full of other fibers. That means, of course, that one errant backhoe may cut the fiber of more than one carrier—and your redundancy that looked perfectly adequate on network diagrams turned out to be in the same plastic conduit all along. How much can you afford to audit that—and how often?

The fact is, rights-of-way converge at common points, such as bridges and tunnels. There are only certain paths that are practical to take between Point A and Point B. Multiple providers may actually take different routes to that common point, but to cross the Mississippi River between Iowa and Illinois, for instance, it's a lot more practical—easier to install and far easier to maintain—if you use an existing bridge than if you go under the riverbed (à la transoceanic cables), especially because navigable rivers such as the Mississippi are periodically dredged to maintain that navigability. All the providers wind up in the same tunnels and on the same bridges. Their rights-of-way led there, and it makes far more sense to use the facility than to create a new one. Thus we have diversity, but only to a point.

Likewise, major highways through the Rocky Mountains and Sierra Nevada are major data highways as well. It simply makes more sense to use a route that not only has been carved for you, but has access maintained year round. As Roeland Meyer noted in a NANOG posting on July 22, 2001, "In short, one can theoretically demand redundant diversity of routes, but may not be able to achieve that goal in practice. Those that have higher expectations need to have those expectations examined."

Required Evacuations

Hazardous materials travel the highways and rails, and even by air, constantly, as well as on pipelines (gas and oil). When a train with tank cars derails, or when a truck is involved in an accident on the freeway or an ordinary street, the hazardous materials may well spill; vapors will go with the wind. Depending on the particular chemical and the atmospheric conditions, a large geographic area may be evacuated. The cleanup effort will often involve large hose-downs from fire-fighting equipment, which causes the problems associated with extremely localized flooding: Circuit conduits may be flooded, power may be shut off as a preventive measure, and so forth. It may be a surprisingly long time before you get back into your facility, depending on the size of the spill and the time required for cleanup.

The possibilities that might cause you to evacuate are endless. A number of carrier hotels were located near enough to the Staples Center in Los Angeles where the Democratic National Convention was held in 2000 that they might have been in the path of demonstrators (one such hotel was only a block from one demonstration rally point). If the situation turned sour, you could face an evacuation and possible fire, looting, or vandalism.

Bad luck can take forms we never imagined. Assume the worst, and use that as a starting point.

Unfortunate Planning

Bad planning happens everywhere, especially when it is bad through failure to consider enough possibilities. It is regrettably too common to find a network design that has optimized the logical topology for administration over service to the network's customers. This bad planning could be yours (local) or it could be someone else's (your provider or possibly the infrastructure). Let's examine these possibilities a little further.

Yours

You have a network that, in addition to its internal information distribution and management functions, supports a large e-business infrastructure. You have a series of Web pages (hundreds at least), devoted to promoting your products, offering the opportunity to purchase them, making available troubleshooting information (a knowledge base), and offering a download site for patches and upgrades to your software. Your Web operation is often a pain to administer, but it is the very public face of your company and a major source of revenue, in the form of orders placed and software downloads purchased.

Prudently, all your Web servers operate from a DMZ, and your intranet is behind a firewall inside the DMZ. Included in your network are your name (DNS) servers; they work through the firewall to obtain name resolution records from other name servers and to offer their information (the mapping of your publicly available names to IP addresses). You also have a primary and a secondary name server (as required by DNS best practices, though you know other companies wink and say they have redundant servers when they really don't).

For simplicity of administration as well as to maintain close control, your primary and secondary DNS servers are located in the same secured server room, on the same subnet behind your newest and best router (in terms of throughput—you don't want name resolution slowness to degrade your network's performance).

Oops. A routine upgrade of the router's software renders it (in the jargon) "toes up." The router has crashed, and you don't even realize it until a pattern in the help desk calls leads someone to realize that no one in your entire company can resolve names. In the meantime, because DNS records expire and yours haven't been refreshed, no one outside your company can reach your publicly available servers, either.

It took Microsoft 22.5 hours to get its DNS back online after just such an incident in January 2001.

If you think it couldn't happen to you, go to www.menandmice.com/dnsplace/healthsurvey.html on page 277 of Appendix A, "References." Men and Mice is a firm that specializes in DNS support. It conducts periodic surveys of the reliability and security of DNS servers, and its results for 2001 were alarming. Among other pieces of bad news, one-quarter of the Fortune 1000 companies—250 major global companies—made exactly the same mistake as Microsoft (all DNS servers on the same subnet), and this was as of September 28, 2001, more than nine months after Microsoft's outage. In the same survey, 29 of the Fortune 1000 had only one authoritative DNS server (though this was a decrease from 40 in January), which means they still have a single network element point of failure for name resolution.

The situation is not much better for U.S. government networks (over 23 percent with all servers on one subnet). Amazingly, while the number of dot-com domains with this condition fell from almost 38 percent in January to 31.5 percent in May, it has since climbed back to over 36 percent as of the end of November 2001. If North America looks bad (and most of the Fortune 1000 have their headquarters in North America), much of the rest of the world does worse. Results ranged from a low of 29.5 percent in Norway to a high of 57.5 percent in the United Kingdom.

> **NAME RESOLUTION—THE QUICK VERSION**
>
> Name resolution is a major topic itself, but here is a quick (and not too dirty) version of how it works. The term *Domain Name Service* or DNS is used somewhat loosely to refer to both the actual resolution of names into IP addresses via an exchange of messages and the system of servers using protocol-formatted messages that performs the service.
>
> Every time a client requests a Web page or sends an email or connects to any other host by name, it is likely that DNS is used. The principal exception occurs when a static table of host names mapped against IP addresses has been created and stored on the local host (the table, an ordinary text file in a certain format, is called a *hosts file*). The client trying to reach the other host sends a name resolution request to its DNS server. If the server has the information on hand (in cache), it answers immediately with a reply containing the IP address for that name.
>
> If the local name server does not already know the address, it asks the question to a name server at a higher level in the DNS structure. If that server does not know, it kicks the request upstairs in the hierarchy, and so on, until the query reaches one of the *root servers*, the master record keepers of name and address mappings for the Internet. If the root server does not know, no one knows, and the reply comes back that the name cannot be resolved.
>
> Primary DNS servers are responsible for *zones*, areas within which they are the authority. Their authority is replicated to the secondary server in their zone (if an organization is too small to support its own secondary name server, it may arrange for another organization's name server to be its secondary, and vice versa). These and subordinate local name servers in large networks also have pointers that help locate other name servers that have more information about addresses outside this network's zone and to whom the authoritative server offers its information. In a multilayered hierarchy that descends from several (replicated) root name servers, the DNS system operates as a logical name resolution network within the Internet.
>
> One final point: While much of the information rarely (if ever) changes, some does, and so there are expirations on records and periodic refreshes of the information. During this refresh process, false or corrupted address records may be distributed. DNS servers should be configured with a detection mechanism for this; unfortunately, some DNS implementations are installed with this feature turned off by default. Especially on a function as critical as DNS, be sure you install the service as you will need it, not necessarily according to the default installation.

In December 2001, the U.S. National Infrastructure Protection Center reminded readers of its Highlights (URL in Appendix A) to provide physical redundancy, preferably in dispersed locations, for name servers. Does your company do this, or have you opted for the easier administration and maintenance approach?

DNS, of course, is not the only service on your network that could suffer from this sort of problem. You should examine your system with regard to all logical services that could be isolated by a single subnet failure, the loss of use (for whatever reason) of one server room or one facility, and so forth. Examine this at both the logical IP layer of your network and at the logical Layer 2 (Ethernet, Frame Relay, ATM connectivity).

When you have done this, start over: Now you must examine your Physical Layer for the same problem. Do not assume that your infrastructure, even at the building level, is immune. In large facilities, for instance, there is a trade-off point at which the cooling required actually becomes more economical with water chillers than with conventional HVAC. This is more likely to be in facilities with large numbers of large systems, such as commercial data storage facilities. Water chillers, of course, depend on a supply of water. What would happen if the water supply is cut off, due to a water main break upstream, perhaps? Does anyone think to have redundant water supplies?

Another infrastructure support to consider is your power supply. Like CDG in Chapter 6, "The Best-Laid Plans," you may not have the right kind of UPS supporting your critical elements. Or, though it seems unlikely, you may have a single point of failure in the power distribution system.

A problem with your building's infrastructure crosses the boundary between your bad design and bad design by others. If you own the building, it is your failure; if you lease the facilities, it is someone else's.

Theirs

On July 18, 2001, a 60-car freight train with 9 cars carrying hazardous materials (including three types of acids, two of which were highly corrosive) entered a 1.7 mile, 100-year old tunnel that runs underneath downtown Baltimore and carries an average of 40 trains per day. The last cars were finally removed from the tunnel several days later, after the fires were out and the hazardous materials neutralized and/or removed so that it was safe to enter the tunnel. I have not found whether a cause for the derailment of four cars (which is not many, really) and subsequent 1000°F fire inside the tunnel has been identified, but that is really not germane. What matters is that at least five major carriers had fiber-optic lines running through the tunnel hung from the wall alongside the rails, and their fibers went down. For one carrier, at least 5 OC-48s and 100 T-3s were affected.

Further, businesses in the area were evacuated (with civil defense alarms wailing) due to the danger from the HAZMAT spill. Complicating matters even more, a major water main broke under the strain, flooding the local streets, and the water could not be shut off initially due to the need to

> ### HOW TO MAKE RATBURGER
>
> Howard Berkowitz, in his *WAN Survival Guide* (John Wiley and Sons, 2000), offers a delightful (and presumably true) story of just such a situation. A national ISP had a Point of Presence (POP) in a building with truly redundant power supplies: Power entered the building in two locations (on separate sides), from two separate grids. The facility had a diesel backup generator. From the electro-mechanical transfer switch governing which of these systems supplied line power, the active power source was routed to redundant UPS.
>
> Notice the article in the previous sentence: From "the" electro-mechanical transfer switch.... In terms of power distribution, it may well not be practical to have redundant power switches; nonetheless, having only one constitutes a single point of failure. When it failed (you knew that was coming), it did so spectacularly. Early one morning (*very* early one morning), the network operators ran to the room in response to a very loud bang. The switch no longer existed; in fact, a significant portion of the wall on which it had been mounted no longer existed. The thick copper electrical feed to the redundant UPS had a brand new gap of several feet. In the midst of the thick smoke, there were bits of something they had difficulty identifying.
>
> The something turned out to be ratburger. A pair of rats had set up housekeeping in the (no doubt) warm, dark, and cozy confines of the switch housing. In the midst of making more rats with whom to share their little love nest, they apparently rolled onto the main power feed. While we can all crack jokes about dying for love, the POP also died, at least for a while.
>
> They could not find an electrician at 3:00 a.m. to replace the UPS feeder and get them connected to at least one of the redundant power supplies. (Could you? Right now?) When the batteries ran down, the POP did, too.

support firefighting efforts in the tunnel with the HAZMAT spill. In some respects, the flooding helped contain the fire's spread because the break occurred about 3 feet above the tunnel, in an area the train had already passed. Baltimore Gas and Electric reported about 1,200 customers downtown, mostly businesses, lost power due to the break and flooding. It almost sounds like the old song: If it weren't for bad luck, I'd have no luck at all....

One provider rerouted its fiber path, taking over 2 miles of fiber and having to cross a dozen streets. In the end, the affected carriers reported running a total of more than 30,000 feet (6 miles) of fiber to reroute their connections. This is not a trivial exercise; in an emergency, fiber can be rolled down the side of the street (inside a conduit, of course), protected by the ever-popular orange barrels or cones, but there is always the risk of an errant vehicle. And the fiber still has to get across intersections. Snaking it through another understreet passage (alongside other buried infrastructure) may reduce the redundancy or pose the hazard of what else runs through those tunnels, such as water mains, which could always break. Underground passages also tend to collect seepage, which must be blown out before the fiber can be run. It is also possible to run the conduit up a utility pole and hang it aerially along the needed path. The second and third options (underground and aerial) take more time than the first (street roll), but they are safer for the fiber, if not for the workers deploying it. At least one carrier was able to expedite the run of a fiber through a pre-planned alternate route (good) and to lease some space in the existing fiber conduit of another (not so good—two carriers in one conduit).

Interestingly, due to the fact that these were mostly data connections, and each carrier did have some rerouting available, no outage report needed to be filed with the FCC. Congestion was a serious problem, though, and several circuits had to be manually rerouted, which took some time, especially considering the number of circuits taken out by the incident.

Worth thinking about is that a train derailment causing significant network disruptions requiring rerouting is reported in the Northeast United States virtually every year.

That was a real event; here is a plausible, but (so far, I hope) hypothetical one. You have connectivity service with a provider (we can pick on CDG again). Providers do not routinely announce every little problem they have to their customers; it might damage their reputation for reliability if they were perceived to be always having these outages when no other provider does. Call it an uncoordinated, but effective, conspiracy of silence. Typically, when an outage is big enough to be noticed, and there has been notice (evidenced by the number of calls coming in wondering what happened to service in this area), the company may post a notice on its Web site and provide information with which the lucky first-tier help desk phone team can respond.

CDG didn't see a need to call around and mention it to its many local subscribers, but it has been affected by a fire in a nearby building. Specifically, the power has gone out in its local area due to the amount of water used to fight the fire. Among other effects, its IP phones are powered off the server, which is drawing from the UPS, so it is making only those calls that are absolutely necessary. This site is handing over operations to its neighbor site and shutting down every piece of networking gear as fast as possible; no one even thinks about DNS, which was optimized for administration rather than continuity. Good DNS administration is not easy, though it obviously can be learned, but CDG has grown fast. It has considered it safer for its quality of service to wait to diversify DNS administration until it has trained its administrators in another location. Therefore, both authoritative servers are not only on the same subnet, they are both in this location. Because the power outage is not expected to last more than another six hours or so, and their regular update to their DNS upstream is not due until well after that, the on-duty operators didn't see a need to force an early update before the handoff; it didn't occur to them in the rush to hand off before the UPS ran out that a customer might be propagating an update to them (customers' networks just don't change that much, after all).

You are just that customer; you recently expanded your network. In this process, you changed the DNS addresses of your name servers, being sure to place them on different subnets as well as in different locations.

When CDG's local site goes down because the batteries have run out; CDG's DNS service goes down as well. The only upstream (in the DNS logical system) that knew about your name server address changes just went off the air before it announced that change to anyone else. Until they come back on the air, the rest of the world is trying to reach you via the old addresses. You have a slightly different flavor of the same problem Microsoft had, and until people start calling you (those who care enough to try rather than simply going to your competitors' sites), you probably don't even know it.

How much is that in lost business? Depending on the degree of reliance in your company on electronic commerce, it could be a lot. And are you big enough and dominant enough in your market (like Microsoft) to weather the attendant bad public relations relatively unscathed, or is the damage more serious? That's an experiment I prefer not to run.

More darkly, consider this possible variation on the problem. You run one of the implementations of DNS that defaults to the security setting being off (as mentioned previously; details are available in CERT-CC Vulnerability Note VU#109475; CERT's URL is in Appendix A). Your primary DNS server's cache (which was replicated to your other servers) has been corrupted, and you realized this only after someone called to warn you that they were redirected to an online pornography site when they tried to reach yours. You flush the cache and send an update to CDG—which, as before, fails to forward the information before losing its connectivity.

Now you not only have lost business, you have a very large public relations headache. Blaming your ISP, even though it was partly at fault, not only sounds pathetically thin, but does not absolve you of any of your blame.

RODENTIA STRIKE AGAIN...AND AGAIN...

Rodents really are kind of cute, with soft fur, big eyes (usually) ... and big pearly whites. In fact, those big teeth are a real problem for networks. Rats are not as cute as hamsters and gerbils, of course, but all members of the rodent family share the characteristic of large incisors that continue to grow their entire life. If the critters don't gnaw, their teeth will literally grow too big to eat with. Researchers count over 2,000 different species around the world that they call FOC—Fond of Cable.

Trenching and tunneling often leave soft earth behind, easy street for moles and woodchucks. Sewers and other understreet conduits are ready-made tunnels, so we may always expect to find rodents, usually rats, inside them and around our networking cables. When they gnaw on the cables so conveniently there, the trouble begins. Even if they don't chew right on through, which actually happens far too often, they may leave a break in the sheathing.

That break allows water in; the expansion due to freezing will eventually break the glass fiber inside, and of course water plus the electricity in copper cables is not a good thing. In those climates warm enough to avoid the freezing problem, the water will still react chemically with the doped glass in the fiber, yellowing it and reducing its transmissivity. The cable industry as a whole cites rodent damage as the primary cause of service interruptions; the category is tied with technical problems for first place in the cause list.

That certainly provides some incentive for countermeasures. AT&T and Qwest both use heavy corrugated steel in their conduits, and so they have fewer than 1 percent of their outages due to rodent damage. Dow Chemical manufactures a specialized steel coating for fibers to prevent damage; some experiments include running a light electrical current through the coating to deter the rodents.

Other choices being tested include a gel inside an outer coating; the gel is laced with capsaicin, the oil that makes hot peppers hot. Because only their teeth generally touch the cable, it disturbs the rodents, but they rarely run off desperately seeking water. Burlington offers a cable sheathing with an extremely bitter taste (human testers typically call it "vile").

Of course, rodents don't just chew on cables. One NANOG post came from an administrator who had his data center disrupted when soaked ceiling tiles crashed down onto the equipment. A contractor had saved a little money by using a plastic pipe for a condensate drain from the HVAC to a janitor's closet 150 feet/46 meters away. Inspection of the plastic piping after the fact revealed literally hundreds of holes chewed into it, creating a home-made soaker hose for the ceiling tiles. In parts of the line, entire sections of 1-2 feet/30-60 centimeters had been eaten away.

Building management blamed the data center company for using a plastic line; the poster blamed the building for having rats. Nobody won.

The timing of your update and your provider's outage was just bad luck. But bad luck, again, is the intersection of preparation and opportunity. In either case, did you multihome (connect via more than one provider)? No. It is more expensive, but it offers redundancy (to include DNS update redundancy, as well as redundant access for you to the Internet and for customers to you). Because you chose not to multihome, did you investigate the details of your provider's DNS service, especially the redundancy and resiliency of it? Apparently not, or else you did not then realize the implications of their single-location, single-subnet configuration.

Had you multihomed, CDG's problem would not have cascaded onto you; in the darker scenario, you could have shortened its period of offensive redirection of traffic intended for you. Had you made an inquiry and made known to your provider that you found its DNS configuration to be a source of concern, it might have rectified the problem sooner. CDG might not, but silence from the customers is an implicit endorsement.

Which should lead you to ask of yourself, what flaws in your configuration have been "endorsed" by the ignorance of your customers? What will their reaction be when their bad luck (in the form of your problem) exposes those flaws?

Unfortunate Implementation

It is possible, of course, to have planned things diligently and well, only to have the implementation of your plans turn out to be unfortunate. This could be because they were only incompletely implemented at the time opportunity knocked. It could also be that compromises were made, for financial reasons. We will get into how to decide on your compromises in Chapter 9, "Preparing for Disaster"; of course, you may well not be the one making the compromise, and the choice may have been taken over your objections. As with the operators at Chernobyl, turning out to be right after the fact is small comfort. Finally, an unfortunate implementation may be due to the nature of the equipment purchased; details of its configuration may defeat your planning, at least in part.

Equipment 1, Plan 0

Serious networking equipment (more than just a local network segmentation router) usually has a multitude of cables connecting a given device to the rest of the world. A tray or panel arrangement is designed into the equipment in order to stow those cables away; not only do you not want a tripping hazard for the human liability issues, you definitely do not want someone jerking your high-speed data cables out of their connectors in the process. T-1 and T-3 cables usually have RJ-series connectors whose pins could be bent. Attaching another connector takes time, assuming you can find the documentation with the correct pinout. Likewise, if the connectors on the ends of an optical fiber are damaged by being jerked, when new ones are mounted the fiber may well need to be smoothed with a little tool to ensure that the glass is clear enough going to and from the laser. This, too, is labor intensive, and the time to reconnect in either case is a service disruption.

The equipment is normally mounted in a rack, so the cables may be run beneath or above, as appropriate to the particular model. In the back or the front of the rack, depending on the site's implementation, panels serve to cross-connect various cables (cable fanout and termination occur here). Cables and fibers are usually run along the sides of the racks when they must change elevation, and they are snugged tightly in place with cable ties to prevent just the kind of snagging and jerking described previously.

And therein can lie the problem. A site with the cables properly stowed according to the vendor's instructions and design may leave the equipment difficult or even impossible to maintain without disrupting service to paying customers. If the cable tray or panel leads cables across other components en route to their stowage, you cannot access those components without loosening and working around the cables (maybe) or detaching them (more likely) while taking careful note (tags are better) of which cable attached to which interface. Setting up duplicate circuits to avoid service disruptions while performing equipment maintenance is hardly economical; maintenance will therefore be a service-disrupting event.

You may think such a scenario unlikely, but it has indeed happened. A client of one of my colleagues had essentially this situation occur. A large and highly reliable main router had redundant power supplies, which worked flawlessly on demand—the primary failed and not one bit of data was dropped as the secondary took up the load. This was the kind of reliability that the equipment was purchased for, as well as its throughput. When the replacement power supply arrived, they discovered that the power supply components of the router lay behind the cable path. The only way to remove and replace the power supply was to detach the data cables—completely defeating the no-service-disruption point of the redundancy.

If equipment is advertised to you as having hot swappable components, that should be verified. But also verify that, *as configured for real use*, you will be able to physically replace those components without disrupting the operation.

Solving the Wrong Problem

This is another instance of bad preparation meeting an opportunity. You may be receiving complaints about slow network performance from users throughout the company; when senior management is among those users, you may actually have the opportunity to do something about it. There are two likely sources of delay in the network and a host of attractive suspects to distract you. Know which problem you need to solve on your network.

Candidates

Among the possible choices to consider are these:

- Hardware location (move the servers closer to the clients).
- Fatter pipes (it must be a bandwidth problem).
- Users clogging the sufficient bandwidth with unnecessary traffic like surfing the Internet, playing online Quake games, and so on.
- Security measures slowing down traffic.
- Lousy protocols and/or operating systems.

Before you actually throw money at the problem, take some time to do some research. You may make your network problems worse rather than better.

For instance, moving the servers closer to the clients may or may not improve survivability from physical disasters; it is unlikely to improve server-client performance. All delay components added together (nodal delay at each hop, transmission delay at each interface, and propagation delay along each link) might change by a fraction of a second; these delays are typically measured in milliseconds, and you will need major changes to create as much as a one-second decrease.

Fatter pipes are always attractive, though as you lease them and purchase or lease the equipment to connect to them, you are likely to see two things occur: The bandwidth that was so ample on its implementation will become just as busy as the older, lesser bandwidth, and performance has not improved.

Misuse of the network may be a factor, but it should not be happening regardless of any performance problems you may be having. Your network use policies should make abundantly clear that you are providing a network for business, not personal, uses. Some policing of this is always necessary to prevent not only pornographic file storage on your servers (it has happened), but also employees setting up their own e-businesses on your servers (that has also happened).

Security measures are a real candidate for two reasons. First, at certain locations (such as every ingress and egress on your network), you should be filtering. That requires an inspection of every packet, which not only takes a tiny bit of time for each packet, but also imposes an extra load on the router's CPU. Moving to a higher-capability router here would help. You could also move to a firewall system that does stateful inspections. Stateful in this context means there is a memory—if two systems have been "cleared" by the firewall to hold a conversation, then all packets identified as being for that conversation are passed through without further inspection. This requires both more sophisticated software and additional memory to hold the lookup tables, but products with stateful inspection capability are available from several vendors.

The second reason security measures may be a candidate for improvement is topological. Security devices tend to be expensive and so are not scattered about the network in profusion. Instead, the network is designed to force traffic through certain choke points, and the security measures are implemented there. This logical design does not optimize throughput; that is a choice that should have been made intelligently. Or, as you may explain to management, it takes a moment longer to unlock your door when you get home, but would you really prefer the alternative? Be careful which problem you choose to solve (or keep your resume current).

Finally, someone may complain that the real problem is lousy protocols and/or operating systems. If we just implemented X instead of Y our traffic would flow faster. Actually, there is merit to this general argument, though the details are incorrect. TCP/IP, like democracy, is the worst possible system, except for all the rest. It is also used on every business-class operating system and application. You'll recall our review in Chapter 3, "Tactics of Mistake," of the Application Layer sending information down the stack, where it is encapsulated and encapsulated some more, until it goes onto the wire or fiber. At every hop some decapsulation is performed, followed by reencapsulation and transmission, until it reaches its destination, where full decapsulation occurs. At each IP hop, at least two quality control checks are performed on each packet—one for the Data Link Layer protocol (Ethernet, frame relay, ATM, etc.) and one for IP. At the destination, another quality control check is performed by TCP on each packet. These are all processor-intensive, which sounds like an argument for closer servers and clients.

The network equipment makers are well aware of this inefficiency, and networking equipment performs these checks in ASICs, very fast and specialized hardware. The network equipment is not the culprit this time. Hosts—PCs and servers—tend to be the bottleneck here. Their processors are performing all these checks in software rather than specialized hardware. A typical Fast Ethernet connection—potentially capable of transferring traffic at 100 Mbps—will actually transfer files at 18-27 Mbps, and the receiving host's CPU is operating at maximum capacity to process this load. It cannot acknowledge receipt of valid packets, a requirement of TCP file transfer, any faster than that. This is one reason why throwing more bandwidth at the problem will not help. Improving the speed of your host CPUs and network interface cards would help because they are the bottleneck for the protocols we all use.

If you get the budget to improve your network, don't be unlucky enough to solve the wrong problem.

Preparing for Disaster

No one ever understood disaster until it came.
Josephine Herbst

The dictionary tells us a disaster is a sudden calamitous event bringing great damage, loss, or destruction. From that, we can infer that if there is not great damage, loss, or destruction, it is not a disaster. Even the best preparations won't take you that far, but they can do a great deal to mitigate the disaster. When it becomes time to recover, that will be easier as well.

Define *Survival*

While it never hurts to turn to the dictionary to be sure of the exact meaning of a word, this time the point is to define *survival* in terms of your business's needs. If your business is in financial services, its needs will be far different from those of a health care organization, which will differ from a news and information service. Once the business's needs are understood, you can then take the process to the next step and define what your network must be able to do to support the business's survival. That set of characteristics and services must survive if your network is to do what it's there for.

What Must Roll Downhill

If the business does not continue, keeping the network going is not important. Senior/executive management must decide at what level they require the business to operate in order to have what they have chosen to define as continuity. This requires an understanding of the mission of the business, which is not necessarily the same thing as the corporate mission statement. Those tend to be lofty goals that should focus the attention of stakeholders, especially management and employees, on what the business will become; they are often stated in terms of social criteria such as integrity and contributions.

The usual business's real mission, as opposed to the for-public-consumption mission statement, is simple: Earn an economic profit for the shareholders.

For a business to earn economic profits, it must, above all, remain in business; continuity does not guarantee economic profits, but a lack of continuity ensures their absence. With that hard truth uppermost in their minds, senior management must define what it will take to continue in business. That depends on the nature of the business; for a news and information service, it is obviously a continuing flow of information to the subscribers—which requires continued operation at a certain minimum level of the information gathering (input), the ability to collate and manipulate that information (product development), and product dissemination (printing, electronic mailings or Web site posting, and so on). A health care organization would have different needs, such as continued operation of the database with patient records, continued ensured privacy and integrity of those records, continued automated monitoring and reporting, especially in such areas as the Intensive Care Unit, and continued information exchange with other facilities and specialists on call.

This kind of continuity does not necessarily mean that all operations continue without so much as missing a beat in every imaginable set of circumstances. Extending the observation of one author in a technical paper on cyber-attack protection, there is a significant difference between the public's (and your customers') perceptions of a disruption caused by a natural disaster versus one from a cyber-attack. In the case of the former, the public and customers expect some difficulties; they look for the company to make a serious effort to carry on, but they will tolerate some glitches in the process. In the case of a disruption due to a cyber-attack,

PROFIT VERSUS PROFIT

Profit is a word whose meaning depends on context. In tax terms, it refers to the money remaining after a series of deductions for the costs of doing business, the inclusion of prior disallowed tax items (such as carryforwards that could not be taken to reduce tax in a prior year), and the legally required tax treatment of certain items (such as depreciation). In an accounting context, it refers to the money left over from revenues after all costs have been paid; it is the change in owners' equity.

The costs reflected for accounting purposes do not necessarily match those taken for tax purposes; depreciation schedules, for instance, are often quite different. As a result, the taxable profit and the profit stated in reports to stockholders may be two very different numbers. For evidence of this, check the figures on major motion pictures, which often show a tax loss while clearly rewarding some investors handsomely. Or look at Enron's books—if you dare.

Taxable profit and accounting profit do share one common concept when it comes to costs, however: Only explicit costs (money spent) are reflected in the calculation. Depreciation is considered an explicit cost because it is intended to capture the diminishment in value of an asset (although there are different versions of how much to depreciate in a given year). There are additional costs that economists consider, and these are implicit costs. The primary implicit cost is the opportunity cost, the best alternative return that could have been earned had the money not been used for this purpose. Often, the risk-free return of U.S. Treasury obligations is taken as the opportunity cost; any other investment carries some risk. For the investor, the Treasury rate of the appropriate duration represents the safe and sure thing; to be worth taking a risk on an alternative investment, the latter must offer to return more. It is not guaranteed to return more, of course. It is always possible that the opportunity cost will be greater than the amount that was actually earned; the technical term for this is investing misjudgment.

If the business's return on a given investment matches the opportunity cost of the alternative investment, the economic profit is zero because that alternative return is an implicit cost to be factored in. In a truly competitive market, no firm will earn economic profits: Competition will drive prices down to a level where those who cannot cover opportunity costs will take their investments to alternatives where they can get the better return. The remaining firms will compete on price and hold it to that level.

To earn a positive economic profit then, a business must provide more value without increasing cost commensurately. Only then will it do better than the best alternative its investors could earn. If it can do this, however, it is attractive to investors and can raise capital inexpensively, possibly lowering its costs; a virtuous circle ensues.

they will perceive the victim as incompetent, with the possible exception of the occurrence of a mass disruption, such as Nimda, described in Chapter 3, "Tactics of Mistake." Even then, however, they will notice that not everyone was a victim; when they separate the wheat from the chaff, victims of cyber-attacks are chaff.

Therefore, the determination of business continuity requirements is a nontrivial exercise. Management must decide first *what* the business requires to continue in operation and next *when* those capabilities must be present; then it can decide on *how* to provide them. Unless the network itself is the product, it is in the process of deciding when and how to provide business continuity that network continuity must be addressed, along with all the other resource continuities. There is one benefit to this sequence: While senior executives will surely scrutinize the costs associated with network continuity, if you can demonstrate that these items are necessary to meet the requirements they laid on your function, they will have to find a way to fund your needs or they will reexamine the requirements.

Survival Requirements

When you address network continuity as part of business continuity, given the constraints and requirements imposed by management, you must categorize your requirements. A simple but effective way to approach this is in terms of necessities, nice-to-haves, and luxuries; apply these categories across your services (hardware and software components), your communications, and your personnel. You will need to do this from two different perspectives: a physical disaster, in which you lose access to a significant facility, probably for longer than is acceptable, and a cyber-attack (whether mere hackers or something more ominous) that leaves your physical plant relatively unscathed but also unusable.

The effects of the two situations on your people are radically different. You will find people stunned, even shocked, by a physical disaster, but they generally pull together fairly quickly and search for ways to get at least some things done; it is a form of coping with a problem whose parameters are clear, even stark. In the event of a cyber-attack, the problem is usually less clearly defined: The boundaries of infection in the network are both vague and fluid, while users are calling for help and updates. Management calls add to the stress on network and system administrators. Rather than pulling together, people are more likely to snap at each other in the stress and frustration of struggling to even define the problem.

The two types of situations then are very different; in the first, though you lose an important piece of your network, the rest is not necessarily at any greater risk than it was before, and so it can not only continue, it can shoulder some of the burden of lost capability. People will try to make that happen. In the latter, the affected network piece is still there, but it is both unusable and a risk to render the rest of your network unusable by contamination. People are the most critical resource, and they are under a less visible, but no less severe, strain as they try to counter an unseen enemy.

Protecting against both types of situation will not be inexpensive, but neither is the alternative.

As we step through this process, we will use the example of a (fictitious) network service provider, STL Network Services, Inc. This firm, headquartered in St. Louis, Missouri, provides complete external connectivity to its business customers. It serves as an ISP, along with providing WAN connectivity between customer sites, managed extranets among its customers, VPNs, and similar high-touch network services. Among its customers is a storage service that uses dedicated lines for FICON (Fibre Connection) and ESCON (Enterprise System Connection) traffic. The storage service supports, among its major clients, a series of databases for hospitals and other medical providers, with high-speed retrieval of patient imagery and records for consultations. STL also provides connectivity for several financial services firms (brokerage offices, banking offices, and so on).

The network service business is highly competitive, and STL's board of directors has decided to distinguish the company by its assurance that it will be there, even in the event of a disaster. STL wants to offer such network continuity as a premium service, available through a service level agreement (SLA). You are charged with determining what network continuity capability STL currently has, what it would need to have, and an economical path from here to there. You will begin by planning such service just for the St. Louis area.

Network Continuity Requirements

Even though this was not the order of your instructions, your first step is to determine what capabilities your network must have in order to ensure continued service to your customers; then you will see what you can currently provide against that standard. This first step requires that you define the mission of your network (see Figure 9.1); while the network's mission is fundamentally to deliver information, be specific about what it must deliver, to whom, and with what degree of responsiveness. With that defined, you can determine what services are required to provide that information and what attributes each service must have.

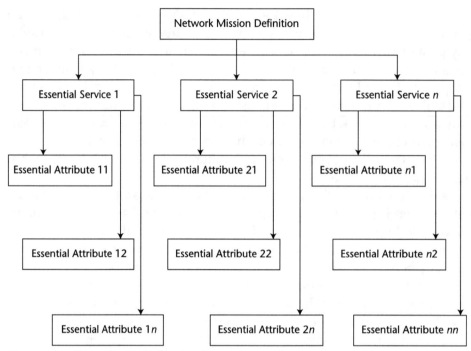

Figure 9.1 Network continuity planning steps: requirements.

In the case of STL, the network's mission definition could be phrased as this:

- Provide network continuity within the STL management network
- Provide the following services to STL customers:
  - ISP services
  - VPN services
  - WAN services
- Provide these internal and customer services on an uninterrupted basis within the STL network
- Provide uninterrupted access to external networks when the traffic's destination lies beyond STL's direct connectivity

Although quite broad, these statements capture the high-level services STL's network must deliver. Notice that first on the list (and really the highest priority) is to maintain internal network management capability. Like the Public Switched Telephone Network (PSTN) and most well-run data networks, STL's highest priority traffic is network management, or OAM, traffic. If you cannot direct changes in the network or identify when and where you need to make changes, the network cannot adapt when problems arise.

This set of high-level services, however, does not yet tell us what exactly the network must provide. For this, you need to decompose the high-level services to their constituent services (see Figure 9.2). This must be taken to the level of fundamental processes on the network elements providing each service. These processes are *what* the network delivers. Each of the fundamental processes will have mission-critical attributes; these are qualitative and quantitative characteristics related to security, safety, reliability, performance, and/or usability. This set of attributes for a process collectively describes *how* it delivers its mission-critical functionality.

This is a place where you could have two groups looking at the problem: like the exercise at the end of Chapter 6, "The Best-Laid Plans," you could have a group of your senior, experienced people, who know the network's little idiosyncrasies from long experience with it, and a group of your sharper, but less experienced people. The second group may be less seasoned, but they are likely to be better aware of the details of implementation (the actual configurations, the gotchas in your OS), and they will see the whole network with a different perspective from those who are perhaps too familiar with it.

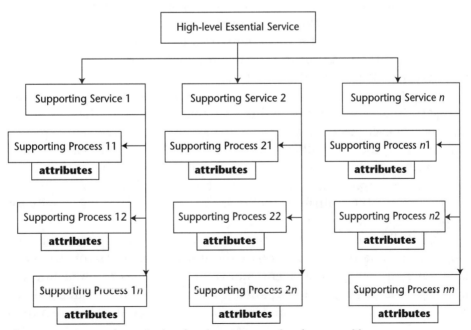

Figure 9.2 Network continuity planning steps: service decomposition.

STL is providing four high-level services: continuity of its network management ability, ISP services, VPN services, and WAN services. Examining each in turn, we look at the network continuity service first. The best way to ensure this is via a separate logical management network; that is, management traffic and production traffic (internal business as well as customer traffic) are segregated from each other. If a problem arises, such as congestion or a need to reroute traffic, it is far more likely to occur among the production traffic. Reconfiguration can be performed because the management traffic takes a different path to the affected device. STL has done this with a port-based management VLAN.

If a separate subsystem cannot be developed strictly for management traffic, a prioritization system may be used. DSCPs would be one method to achieve this; network management traffic would have a DSCP that is given the highest priority on every link. WAN networks achieve the same thing with priority egress queues for management traffic.

The second service, ISP services to customers, has a series of subordinate services; the exact set may vary from one network service provider to another. Individual customers may select some or all of the service set offered. Three services that are virtually certain to be present are Internet connectivity, DHCP address assignment for businesses that do not want to administer it themselves, and DNS. For those wishing to use private address space, NAT (Network Address Translation) can be offered; this is an address management feature that also has security benefits. A service that customers probably do not realize their ISP provides is address advertisement to the Internet for those networks not using RFC 1918 private address space (and NAT). And, finally, a network service provider may well also provide a Web hosting service.

Each of the ISP services STL provides has its own attributes that are necessary to service continuity. Internet connectivity will have a set of attributes for each customer according to the contracted agreement, typically an SLA. The attributes will include a guaranteed minimum bandwidth, a bandwidth expansion the customer may burst to on an infrequent basis (with the understanding that too much traffic too often is subject to discard), and a reliability attribute (definition of an outage and the penalty for the provider in such cases).

DHCP will have an address scope from which host addresses may be assigned, a standard lease duration for those addresses, and so on. DNS will have reliability and responsiveness attributes (changes to name-address mappings will be propagated upstream within a certain time interval, for example), along with zone administration performance attributes. Address advertisement to the Internet will have an attribute of how the public addresses are advertised (separately, or subsumed in the overall address block assigned to the ISP). Web hosting will have attributes for responsiveness to customer change requests, mail rerouting (such as mail addressed to webmaster@xyz.com), site maintenance, site security, and more.

PRIVATE ADDRESSES AND NAT

When the IPv4 addressing scheme was developed, no one imagined an Internet like that which has developed. As a result, addresses were not allocated efficiently (in terms of the numbering scheme or in terms of geographic distribution). For instance, Nortel Networks has a Class A address block, the largest original size, which has the capability to uniquely identify 2^{24} hosts—16,777,216 unique hosts. Even before the radical downsizing it undertook as a result of the telecom market's collapse, it was not that big. But Nortel is not alone; Apple Computer also has a Class A address block, and so do a number of other early adopters of IP.

One of the methods developed to extend the life of version 4's addressing was the designation of Private Address Space in RFC 1918. Certain addresses (10.x.x.x, 172.16.0.0-172.16.31.255, and 192.168.x.x) are designated as private address blocks that should never be advertised beyond their local network. This means that they may be used and reused any number of times—but no one can reach a host on one of those address blocks from outside because the host's address is not known to the outside world. To reach a host from the outside world, it must have an address the public can reach—a public address (public addresses are, loosely speaking, all other valid IPv4 addresses).

If a host has a private-space IP address, in order to be reachable from the outside (to include receiving email and the reply from a Web request), a service called Network Address Translation can be used. NAT translates the private address of a host on the inside to one that is acceptable outside (a public address). Because not every host is communicating externally at the same time, fewer public addresses need be available to the router performing NAT. NAT can also track host-to-host conversations via the TCP or UDP port in use; it can therefore use the same public address for multiple conversations, even if they involve several separate hosts inside the private address block. This further conserves the public address space.

An ISP can assign the same private address blocks to multiple customers; because the outside world never sees their addresses, they are somewhat protected from network attacks (somewhat because it depends on the network security consciousness of their ISP). A tracking function to map conversations to the appropriate port ensures that the data is not sent to the wrong network, even though the IP addresses are the same.

These attributes may seem more routine performance characteristics than necessary attributes for network continuity. But because what STL is going to market is the continuity of the networking service it provides, it must be able to continue those characteristics the customer believes are important (such as those it has contracted for). What is more, in the event of a disaster, like the situation when the World Trade Center was destroyed, people may be able to communicate most reliably via email and companies may be able to disseminate information faster and to a far larger audience via their Web pages. Therefore, the timeliness of updates in response to customer requests must be considered in this example.

For the high-level VPN service, the essential services list is shorter: tunneling and encryption support. Not every customer uses encryption; some are satisfied with IP-in-IP encapsulation. For those who wish to encrypt but lack the in-house expertise, STL offers a service of setting up and maintaining IPSec at the tunnel endpoints. This requires customer premises equipment (CPE) that the customers lease from STL. Attributes in both cases emphasize the security of the customer's traffic (confidentiality must be maintained), the reliability of the network service (documented in the SLA), and the performance—security must not seriously delay throughput. (Note: The patient information transmitted to/from the storage service is required by federal law in the United States to be encrypted.)

The WAN service is a series of point-to-point links between offices of customers, between them and parent organizations (in the case of the financial services firms), and between customers' offices and their extranet partners. STL also offers an extranet administration service. The point-to-point service has no subordinate services; essential attributes focus on reliability and performance. The extranet administration service has characteristics of both VPNs and point-to-point links; both sets of attributes must be maintained.

Once you know the services and the attributes necessary to provide them at the required level for continuity, you must determine how those services are currently provided. You must trace them through your network, from the requesting host to the responding host *and back again*. Remember that, with TCP/IP traffic, the response traffic route need not be at all the same as the request traffic route, other than the two endpoints. Your routing protocols, management policies (via iBGP, for instance), and any traffic engineering you use will direct this. During the service trace, you may find critical elements (nodes or links) where redundancy is not currently available (see Figure 9.3). Those services are vulnerable to a failure in such a single point.

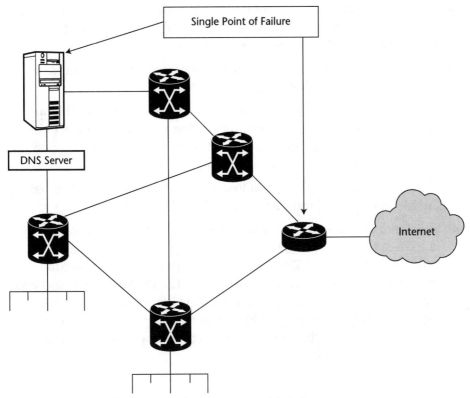

Figure 9.3 Network continuity planning steps: critical elements.

In the case of STL, the following critical elements were located:

- The management network is a subnet of STL's overall internal IP network; it does not use a separate IP addressing scheme. It is thus vulnerable to manipulation from any host inside STL. While STL's intranet is well protected from external cyber-threats, internal cyber-threats have no blockages, and, because the external protection cannot be guaranteed, internal protection is required as well. One dual-homed workstation is currently directly connected to one routing switch; any intranet host may telnet to that workstation and manage the routers and switches from there. This was done for convenience of network servicing from administrators' desktops or any other handy host where they happened to be.

- All OS installations throughout STL's intranet were default configurations. Many unneeded services are available on most hosts, including networking devices and servers.
- Two of the extranets managed by STL have multiple instances of single-node and single-link connectivity.
- The WAN service in three locations is accessed via the local loop; all three of those locations terminate at the same CO. If the CO is lost, those three locations have no connectivity, including a possible backup service via dialup connection.
- Physical redundancy has been lost in two areas: the bridge across the Mississippi to Illinois and along the I-70 right-of-way.
- The connections between the storage service and the medical services use IPSec for encryption, but it is an older version of IPSec. While compliant with RFC 1826, that did not require replay protection (packets may be intercepted by a man-in-the-middle attack and then replayed to the original destination). This is actually not STL's problem, as the IPSec is managed by the storage service. STL will notify the service of the vulnerability; if STL could find it, presumably a hacker could as well.

Threat Analysis

Once the network's critical services and the critical elements needed to support them have been defined, it is time to analyze what could go wrong. Once you know that, you can assess STL's current ability to provide continuity. In this "what could possibly go wrong" analysis, while you don't want to decide by committee, where everyone (or at least a strong plurality) must agree, you do want multiple viewpoints. Threats will come down to a physical loss of use of some sort—a natural or an unnatural disaster—of varying duration and a cyber-attack of some sort. Each of these covers a range of possibilities, and you may wish to shuffle your previous teams into two groups with a mix of experience levels and assign one group to each area.

Physical Threats

You may face a range of potential physical problems in your operating locations. You have the natural hazards endemic to the region—hurricane, tornado, severe winter storms, flooding, earthquake, volcano—and you have the effects of a human action (including an attack), which cause the loss of use of a facility—local flooding, fire, power outages/unreliability, or complete loss of connectivity until repairs can be made (which may exceed 24 hours). For each location from which your network operates, you must assess the threats that may occur, the duration over which they are likely to cost you the use of the facility, and the likelihood of their occurrence.

Your team's initial work will be from your documentation; however, if at all possible, you should try to conduct a physical audit of links and hardware/software actually in service at each location—remember Verizon's discovery of discrepancies as a result of their outage, discussed in Chapter 8, "Unnatural Disasters (Unintentional)." The equipment audit does not have to be done entirely by the team; it can be a general exercise at each location. Regarding links, confirming physical redundancy is difficult and takes considerable time. If you have not built an audit requirement (preferably a semi-annual requirement) into your SLA with the carriers from whom you obtain service, it may not be possible. You can, however, insist on it when the time comes to renew your service contract. In the aftermath of the September 11, 2001, terrorist attacks and the enormous service disruptions that resulted, many companies are conducting just such reviews. The executive who oversees telecom connectivity for Lufthansa traded his suit for jeans and work boots to crawl through train tunnels and verify physical diversity post-September 11. His airline had previously had what it thought were redundant circuits to its two call centers; his job that day was to make sure his new redundant fiber lines did not go through the same conduit.

EFFICIENT VERSUS EFFECTIVE

Many years ago I read a comment in a book that I believe was by Peter Drucker. Because he has been writing books for 60 years (literally) and has dozens in print, I have not been able to chase down the exact reference. Nonetheless, the idea is important (and if it came from another source, *mea culpa*): There is a difference between being efficient and being effective.

Efficient means doing something in the most economical fashion: doing the thing right.

Effective means obtaining the desired goal: doing the right thing.

It is entirely possible to be efficient without being effective or to be effective without being efficient. Both waste effort, but the former wastes it uselessly, while the latter at least accomplishes what was needed.

You will almost certainly have limited resources with which to make your network survivable. Being both effective and efficient will be crucial, but do not sacrifice effectiveness for efficiency.

When the team considers your network in this light, they will look at each location separately. You will need to direct them back to the questions they discussed from Chapter 6; you may already have the solutions to the vulnerabilities they will find in your network. If Site A goes down, for instance, it may be that the critical services from it can be filled by Sites B and C, possibly operating together and possibly each providing a subordinate service. In this case, the higher-level service may be less efficient, but as long as it is effective, you do not have a vulnerability.

Once your team has considered each individual site's physical vulnerabilities and then reconsidered them in light of potential offsets from other sites, you have your current physical vulnerability status.

Cyber-Threats

Your cyber-vulnerabilities team has in many respects a more difficult job. They can approach this from an outside-in viewpoint or an inside-out viewpoint. Outside-in assumes they are on the outside, looking for ways to get in; where are the holes in the defenses? Is there an ingress not protected by anti-spoofing or other filtering? Does product test literature indicate that a particular router or switch is vulnerable to a SYN flood or a flood of malformed UDP packets? (Note: The vulnerability may not be in the hardware, but rather in the software version currently running.) Does a war dialer indicate an unauthorized modem somewhere on the network to let someone in?

The alternative approach, inside-out, assumes the cyber problem source is on the inside. A good place to start in this approach is with the SANS/FBI Top Twenty vulnerabilities discussed in Chapter 4, "Murphy's Revenge." This is a living list, so be sure that your team operates from the most current one. If your organization is like most, an audit here will uncover a large variety of anomalies, not all of which are necessarily problems.

For instance, one company I have worked with has a measured migration from the Windows 95 system to Windows 2000 in progress (over literally tens of thousands of hosts). At the same time, the hardware leasing program for its mobile engineers has equipped a number of them with laptops that have a USB port but no serial port; the laptops came configured with Windows 95 (the default in CY2000 for this particular firm). The problem is that their job focuses on maintaining and upgrading switches and routers that have a serial interface. Windows 95 does not recognize a USB port, so these engineers bought USB-serial port adapters and loaded Windows 98 on their laptops. They can do their job, and Windows 98 is not significantly different from Windows 95 for their other job-related activities. But the upgrade was entirely unofficial and unauthorized; suffice it to say that, having made that change with no consequences, the OS was by no means the only unauthorized software on their systems. Being engineers, they went out and got what they thought they needed and (of course) what they wanted. You will probably find the most anomalies in the most technically adept sections of your user population. Those may or may not be the most dangerous vulnerabilities.

The threats to be assessed run the gamut from voyeurs to highly sophisticated industrial intelligence gathering and sabotage and extortion. Problems that leave you vulnerable to a cyber-attack are less likely to be mitigated by other features of your network, though the possibility must be considered. Once your team has gone through its assessment and evaluated the network as a whole as well as the individual parts, you have your current cyber-vulnerability status.

When the two analyses have been combined, you have an assessment of where your network is weak and where you cannot guarantee continued support to the business. Of course, that is its condition as of the time the report was developed; its condition will change and evolve. If nothing else, you may lose physical diversity of the links to which you connect if rerouting results in the single-conduit problem. Like the telecom best practices cited by Verizon in its outage report, this evaluation should be repeated at regular intervals. Telecom best practices say it should be done semiannually; if your network's performance is indeed critical to the business's continuity, that is your requirement as well.

The combined analysis for STL revealed the following physical threats and cyber-threats in the St. Louis area:

- Physical threats
  - Flooding from either of two major rivers (Mississippi and/or Missouri); the Illinois River, draining a third basin, empties into the Mississippi a few miles/kilometers upstream.
  - Tornado/severe windstorm.
  - Severe winter storm.
  - Earthquake (low probability, but very high destructiveness).
  - Local flooding and/or power outages due to accidents.
  - Connectivity loss due to loss of physical redundancy (bridge, highway, single-homing to one CO).

- Cyber-threats
  - Ordinary hacker probes and reconnaissance.
  - DoS/DDoS attacks on the STL intranet.
  - DoS/DDoS attacks launched from the STL intranet.
  - DoS/DDoS attacks launched from customer networks.
  - Web page redirection:
    - Hyperlink corruption (content).
    - DNS corruption.
  - Intrusion and information collection on high-value networks:
    - Medical records (for blackmail, extortion, etc.).
    - Financial records (for embezzlement, extortion, etc.).

Operational Analysis

When you know the weaknesses of your network and the threats you face that may expose your weaknesses, you still do not yet understand the vulnerability of your network to disaster. The next piece of the picture is your people. You must undertake a realistic analysis of whether your people could respond to a physical threat and/or a cyber-threat. The two should be studied separately; when combined, you may find you have critical human elements: a few highly-skilled people who would be necessary to counter either. If you unfortunately must face both at once, how much could those individuals accomplish?

In addition to the skills possessed by your people, you should take some measure (though it will probably be more of an ad hoc judgment) of their readiness to respond. How aware are they of the outside world—if warnings were given of a pending natural disaster, would they notice? Do they know the number to call for adverse weather advisories by your firm (do not come into work today)? Do personnel with critical skill sets maintain a 24/7/365 responsiveness (probably shared—an on-call rotation)? Have you checked to see that the on-call person can be reached and would respond?

Does someone review your logs? How often? Intrusion detection often begins with an anomaly in a log—are they familiar enough with the usual content that they would recognize a nonroutine entry? Does anyone maintain familiarity with cyber-defense Web sites, mailing lists, and newsgroups? Do you have any sort of ongoing training or personnel development program to keep your people abreast of technological changes in networking? Of the changes in the cyber-threat?

Finally, there is the matter of initiative. As Clausewitz noted in *On War*, no plan survives contact with the enemy. When things go bad, any plan you have will be a point of departure. How willing are your people to depart from the ordinary? Do they show initiative routinely, or does your company require adherence to procedures? If the former, your network may have some interesting though not well-documented features, the result of people doing what it takes to make things work. This adds ammunition to the semi-annual audit requirement, too; updating and validating your documentation will uncover these features and may prevent you from developing an unexpected single point of failure. If the latter, your procedures had best be all inclusive without being so voluminous that the correct procedure cannot be located quickly when needed.

At this point, you finally know where you are today. You know your network's weaknesses, you know the likely ways in which those weaknesses may become exposed, and you have at least a rough assessment of your personnel's ability to respond to a situation. Now it is time to decide how to fix the weaknesses and, if possible, increase the strengths.

Survival Planning

Your planning to make your network (more) survivable has two steps: fixing those problems you can and remedying those you can't. Implicitly, there is another evaluation that you will undertake here: deciding which, if any, of the problems are not worth fixing or remedying. The solution may be too expensive for the value gained, or possibly the solution is indeterminate and so its costs cannot be reliably estimated in order to assess its value.

Fixes

Fixes are those actions possible with the resources already on hand or already programmed to be on hand shortly. Fixes also come in two basic categories: those that have low cost and are relatively quick to implement, and longer-term, more expensive changes (because they involve reordering existing major activities). The latter are generally further reaching in their scope as well. A quick fix is relative, of course; short-sighted thinking probably helped cause the problem in the first place, so you don't want to be too quick in the fix. You want to repair this problem without causing another, especially one that you don't realize is there until your semi-annual review uncovers it. It is worth taking a little time and deciding how this little problem fits into the greater scheme of things, then create a fix that is complementary to that scheme instead of one that undermines it in a new and creative fashion.

Quick fixes generally do not require any new purchases; they are more likely to be rearrangements or reconfigurations of existing assets. That hardly makes them costless, however. Their principal cost comes from the redesign and reconfiguration work by staff already on hand. To get it done right this time (effective instead of only efficient), you may need to invest in some training or time spent studying the existing software/hardware documentation or materials available via the Internet (*caveat emptor*).

Longer-term fixes are those that, again, do not necessarily require you to purchase any new equipment beyond what was already scheduled; the fix lies in how your existing and new equipment are used in the network. There may or may not be modest increases in cost of the new equipment, if you choose to moderate a weakness via more suitable software or a more capable component. You may choose, for instance, to buy new routing switches as planned, but you may choose a faster CPU option in order to add more filtering. You can thus upgrade your network's security and traffic management significantly with a relatively minor increase in cost over what you were going to do anyway. You may recycle the equipment being replaced and use it to logically partition the network further, improving performance by reducing the number of users sharing a given section of bandwidth.

Because the longer-term fixes will be used to evolve your network in the direction of more survivability, it is important to consider these even more carefully. These tend to become architectural changes rather than merely reconfigurations within your existing architecture. They will involve network design and configuration evolution, which will require time to plan and review for consistency and migration problems. It is especially important to have determined the goal architecture first (a part of the remedy phase) in order to be sure that each fundamental change from here forward contributes to the migration toward that goal.

In STL's situation, two immediate quick fixes are apparent: reconfigure the router and switch management network and remove unneeded services and processes from all hosts. The management network should remain a VLAN, and port-based is actually desirable, but it should be on a separate IP addressing scheme, and it should not be readily accessible from within the intranet. If STL's intranet, for instance, is a 172.16.x.x, then the management VLAN should be 192.168.1.x.

The current management scheme, using a portion of STL's general address pool, means that management traffic must cross the backplane/fabric of the switch in order for an egress port decision to be made. If the network suffers serious congestion, such as a DoS/DDoS attack, even if the management traffic has a high priority, it will have difficulty getting through. By using a separate addressing scheme, a separate process within the switch can handle the traffic, and it need never enter the backplane/fabric and be subjected to the delays of congestion there.

The single workstation directly connected to the management port of one switch is a single point of failure once a separate addressing scheme is implemented (previously, because the routers and switches were on the same network as everyone else, they were reachable by everyone, from everywhere). A second directly connected host should be available, at another site. Neither host should be accessible from the intranet. If dual-homing is considered necessary on these hosts to enable telnet in from other hosts, strong access control to directly connected workstations must be implemented (and strong physical access control should be implemented at all locations that can be used to manage the network).

The second quick fix, undoing the effects of default and unauthorized OS and software installation, will be grievously unpopular. Too bad. Ideally, files would be backed up from each host and, during a maintenance window, they would be reloaded from scratch with a tightly configured installation (only those services and applications necessary to its actual business function being installed). All current patches would be applied. On PCs, a configuration-monitoring client would be installed, allowing software tracking to occur, and regularly scheduled hard drive backups would be performed (which would necessitate leaving the system on during the backup window, probably one night per week). Finally, a client that enabled patches and upgrades to be pushed to the host would be installed. Tampering with the host's configuration would lead to a warning on the first offense and termination for cause on the second.

But we don't live in an ideal world, and neither does STL. Configuration control is an ongoing battle, in which the best we can realistically hope for is a draw (famously compared to kissing one's sister, but, in this case, preferable to the alternative). What can be done is to require, at every location, removal of unneeded services and software (instant messaging software, for instance, is even more popular with hackers than with users—it provides an easy entrance into the OS). Every host, servers as well as PCs, should have the latest software patches applied. This should not be a task beyond the local IS team, though it will take some extra time and will generate friction among users. There is a straightforward preventative measure for the latter: Publish throughout the company a computer and software use policy that clearly states the business purpose of the equipment and establishes that configuration control belongs to IS instead of the user. It also establishes clear consequences for violations.

Perform the reconfiguration on senior management's systems first, and make that known as well.

Longer-term fixes for STL involve adding links and switches where customers have only one access CPE and one link (if they are to have the continuity service). They will also involve running redundant local loop connections to the three customers who are single-homed to the one CO; the new connections should enter the customers' premises on different building sides and should terminate in a facility some distance away from the current CO to avoid a common disaster eliminating service from both. Finally, STL must establish a requirement in its SLA with its providers to ensure diverse routing and the right to audit the route. In all likelihood, the only recourse STL will be able to receive if its provider loses diversity is a credit on STL's bill. This is because its provider is purchasing service from a larger one, which may be a major carrier or which may be an intermediary before the major carrier. Until the major carriers and backbone providers offer contracts with teeth in the SLA, those who resell connectivity from them cannot offer what they do not have.

Remedies

For those problems that you cannot solve with reasonably quick fixes and are not yet solved by your network's evolution, you need a remedy. For instance, you probably cannot ensure that your redundant data routes will truly remain diverse (recall Chapter 8), if indeed they ever were. Likewise, it may not be practical, even in your long-term plan, to have more than one facility in one major metropolitan area. Your business's functions may therefore remain subject to a single point of failure based on the single location.

Assuming that you do not wish to be at the mercy of a single failure point, you may choose to contract for backup facilities and equipment (servers, desktops, all configured with the appropriate—or at least usable—software). This is akin to buying an insurance policy; you hope you never need it, but you prefer the premium to the alternative. Your other choice is to restructure your network so as not to have any single point of failure, be it in equipment, locations, communications, or personnel.

This is more expensive to operate, both because you have more than you need to run things on a daily basis and because you must keep all the (redundant) information sets the same. A user querying a database should get the same response no matter which particular instance of the database responds. If the database is updated only at night (say, from 3:05 A.M. to 5:30 A.M.), as some insurers do with their coverage databases, then maintaining consistency is relatively easy; updates are posted to all applicable servers simultaneously. This is less expensive in communications terms because the circuits you lease (probably on a full-time basis) are likely to be underutilized during this period.

If the database is real-time, such as that required for transactions in the financial markets, then there must be constant replication between the instances of the databases. This generally requires a dedicated set of connections (redundant, of course, with routing as diverse as you can obtain). Integrity and consistency checks should be ongoing; consider the penalties that could ensue if a legal claim could be proved that your service of a transaction was in error according to your own information at the time.

The choice of the degree of network continuity, and therefore the means by which you achieve it, must be taken at the senior management/executive level. It will be a major cost; how you evaluate the alternatives and present the costs and benefits is the subject of Chapter 11, "The Business Case." It remains possible that some services of the network will continue to be less survivable than they might be, if the cost to change that exceeds the expected benefit gained.

You may not get what you wish for. You may not be able to afford what you need. You will simply have to do the best you can—and document what you can and cannot do, ensuring that management is aware of the limits on your capabilities.

Procedures

Finally, the network is administered by people. Your network continuity plan should include an explicit plan for operations as well as connectivity and information replication. In the operational portion of the plan, you should be clear as to who has the authority in an emergency to do what. Included in this must be that, if a site loses connectivity with headquarters and no other site has connectivity, headquarters is presumed down (at least temporarily), and its networking authorities devolve to specified positions (Note: not necessarily its business authorities). It is important to specify positions rather than individuals, who may change jobs within the company or even leave. Likewise, if you have a 24/7/365 operation, you know certain positions will be filled with a person when things fall apart. The positions or sites in whom authority resides must have designated successors.

There must be procedures for whom to notify (customers, partners, vendors—especially communications service providers because on-call bandwidth may need to be brought online). The authority to represent the company in these notifications must be explicit.

Finally, recall one of the lessons from natural disasters, as to why the National Guard or militia often succeeds in managing things until the local government can reconstitute the normal operations: They practice what they must do. If the first time you try your continuity plan is when the debris is raining down from buildings collapsing in an earthquake, you may find the plan has gaps or even gotchas. You will need to try it out multiple times to get all the wrinkles out; among the wrinkles will be people knowing the plan exists, knowing how to access it, being at least roughly familiar with its contents, understanding that their jobs will change from the routine, and so on. Rather than a full-blown, pull-the-plug exercise every time, most of the time you can conduct what the military calls a command post exercise (CPX): The only people exercised are the decision makers, and all the people they interact with are simulated by an exercise team.

A CPX also allows you to vary the scenario inexpensively. It provides an opportunity to exercise your network managers on handling an infection or other cyber-attack; rather than actually seeing log entries from simulated events, the exercise team can provide dummy inputs. Symptoms of the intrusion can develop in the order the team being practiced thinks to look (ask) for them. Training exercise development will also feed back into your plan. As the team creating the exercise looks for things to test, they will examine your continuity plan with an eye toward what could go wrong. The things they find will surprise you.

Creating and undergoing CPXs are an excellent ways to develop your people. They offer training that is more interesting than a class or reading a book, team skills that will be needed in a disaster are developed (and teams that cannot get along even in Paradise can be identified and reorganized), and they offer the opportunity to earn recognition, certainly from their peers, possibly from senior management.

And the practice will pay off when disaster strikes.

You will recall the charge given in our simulation was to determine what network continuity capability STL currently has, what it would need to have, and an economical path from here to there. Having taken the second task first, in order to establish a standard against which current capability can be assessed, it is now time to measure what we have today against that standard.

Survivability Today

If you needed it right now, what is the current network continuity plan? If you do not have one, you are hardly alone. But in the light of the magnitude of the communications disruptions among the densest concentration of networked businesses in the world, Lower Manhattan, and how far the ripple effects of that disaster spread, many companies are reassessing what they could do if such an event occurred in their network neighborhood.

If you have a business continuity plan, does it address only the succession of the senior leadership and implicitly assume (without even stating so) that the information their successors will need to make decisions is not only extant, but readily available to them when and where they need it? If it does, you need network continuity or else the business continuity will fail before it starts.

You have three fundamental choices in network continuity:

- Trust in the dedication and ingenuity of your middle managers and line employees to find a way to handle things when disaster strikes (sometimes called betting on the come).
- Obtain access to on-call facilities that your planning indicates should be sufficient to bridge the gap until a more permanent solution can be obtained.
- Design your network so that no one location, if lost, whether due to physical disaster or quarantine to control the effects of a cyber-attack, has major effects.

The first choice is where many companies are today. The second is the approach taken by the New York Board of Trade, which had on-call facilities with Comdisco (which has since been acquired by SunGard). The third choice in some respects is the most expensive, but it is also the most reliable if properly done (recall that, when the WTC attack occurred, Comdisco had been in Chapter 11 bankruptcy for a few months—it might not have been able to deliver if too many corners had been cut along the way; that it could and did deliver speaks volumes).

As part of your reality check, recall some of the lessons learned from the success stories in the WTC disaster:

- They all had a disaster recovery plan in place, with known facilities already available.
- The senior leadership did not hesitate to activate the plan.
- They had to adapt because the plan was less than perfect (no plan ever survives contact with the enemy).

And the lessons learned from those who tried gallantly, but could not succeed, include these:

- Plans not implemented are not a help; money must be allocated and spent.
- Plans must include the loss of the location from which the leaders will direct recovery.
- Preparation for other contingencies can help when the current plan falls apart.

While senior management will ultimately pass judgment on how much to spend on network continuity—and the level of spending will have a considerable impact on the type of plan you can implement—it will largely be up to the senior networking people to advise on what should be done, the best way to do it, and what can be done within the given budget.

Whichever choice you take—on-call leased facilities or intranet design—you must start from where you are today. If you have a network continuity plan, reevaluate it in light of the evolving cyber-threat (better attack tools + fewer skills required = more attacks, which are each more dangerous) and of the physical threat, which most network administrators have not considered. If you do not have one, start the process to make one.

Don't Get Too Close

Redundant facilities were originally a duplicate set of hardware and software. The two sets were kept close together to facilitate administration and maintenance, and because no one thought distance mattered. After the advent of serious hacking attacks, some firms began locating their redundant servers on different subnets; physical proximity still tended to be the norm. This, again, made software migrations and maintenance (such as patch administration) simpler, since only one group of qualified people was needed.

Some firms, observing that something could happen to a given building, began to locate the redundant equipment in a different site (building, anyway). Due to the expense of connecting the two (which increases with distance) and the continuing problem of administration (as software became more sophisticated at the same time qualified and competent IT personnel became harder to find), the second site still tended to be close. Some firms in the WTC had their secondary site on a different floor of the same tower, while others had the secondary site in the other tower. Neither was far enough away.

In fact, how far away is far enough is being reconsidered in light of the terror threat evolving in the way it has. One rule of thumb used to be 6 miles/10 kilometers. However, many facilities that distance apart were both served by the same CO, and so were both off-line due to the events of September 11, 2001. Sufficient physical distance that the loss of site one will not affect site two must be the minimum separation required. When you assess that, you must look at the local topography and the local infrastructure—all utilities and communications, as well as transportation.

If, for instance, you will need to move certain people from the primary to the secondary site in order to operate (certainly true in the case of leased facilities), what will happen to transportation routes in the event of a physical disaster that forces you to move? Will transportation to the alternate location be possible? Who would have expected all flights in the United States to be grounded, much less for as long as they were? More prosaically, travel from New York into New Jersey depended on bridges, tunnels, and ferries, all of which were subject to shutdown in the event of a follow-on threat. You know the ordinary choke points for traffic and transportation in your operating locations; when disaster strikes, consider what will happen to them.

As you review utilities, think about California's rolling blackouts and brownouts in 2000 and 2001. While one may argue the role the state's partial deregulation of the power industry played, add that scenario to your list of considerations. A concern of U.S. federal anti-terrorism planning is an attack on infrastructure, such as power generation facilities. If your primary and secondary location both draw from the same grid, you have a single point of failure in the power supply. While wheeling would bring in power from other regions, there may well not be enough excess power to completely replace a lost major facility, or there may not be sufficient connections available to transfer that much power in (power lines face the same rights-of-way issues as network providers, with the addition of public concerns over the unsightly appearance of major transmission lines, and periodic fears of health hazards despite the continuing lack of evidence). This power supply crunch would be worst during the summer air conditioning season.

Talk Is Cheap

But talk is not cheap when people do not know how to reach you. If your contacts with your customers, your partners, and your vendors are principally by hard copy, what will happen to communications with these people when you lose a facility? Will paper mail be forwarded (how will postal authorities know) and how long will it take? Will faxes simply pile up at the old location? In the more common case of electronic communication, who has the responsibility to notify these interested parties of the new communications routings required? If you run a mail server for 800 accounts, and you must transfer those accounts to another server, this will take some time. Until the email system is updated, will you be as lucky as *The Wall Street Journal*, whose server ran off the generator for three days because the data center was on the opposite side of the building from the destruction?

If someone has the responsibility to notify key business contacts, does that person have the authority to do so on his/her own initiative, or must the action be authorized by someone higher up in corporate management? Consider the likelihood that such an authorizing person may be incommunicado; recall the only means of communication for those evacuating the World Trade Center was the BlackBerry system—the PSTN was already out and the cellular system was already overloaded.

As you consider communications with people in the event of a disaster, consider also your communications with employees. You must have a part of your plan that addresses how to inform them of where to go to work. If you have an on-call lease arrangement for backup, it is entirely likely that very few of the people who will need to go there even know of the arrangement, much less the address of the alternate facility, what the available transportation routes may be to get there, or when they should be there.

Another aspect of your communications when operating from a backup facility is the sufficiency of communications. Remember Lehman's experience—they increased the bandwidth to their surviving data center in New Jersey, since it would be handling information from all 45 branch offices instead of sharing that with the destroyed center. They also relocated a portion of their communications-intensive workforce to a facility which had insufficient connectivity, data and voice, and it took some time, in the middle of a massive communications disruption, to rectify that. Know at least an estimate of the communications load at the alternate facility, and be sure you include in your plan sufficient bandwidth for both the data and the voice communications you will need there.

If you are migrating to an IP phone system, what does its service trace look like? Be especially aware of the redundancies of this system, consider how the phones are powered (a PSTN phone receives DC power via the circuit, which is why you still have a dial tone and can call the power company to notify them of an outage), and consider the effects of network congestion. In the event of a physical disaster, more and more people are turning to the Internet to stay abreast of the news; data network congestion was heavy at news-related sites on September 11. But there were PSTN congestion problems that day as well (witness AT&T's outage report), so IP phones may or may not have better connectivity than a PBX into the PSTN.

IP phones, properly deployed, should not be vulnerable to viruses (some vendors' implementations do use an NT server, but if the server is strictly configured and has all current patches, it is generally not vulnerable—you do have to stay on top of the security topic, though, just as you should for all your systems). Congestion in your network due to a worm or a DoS/DDoS attack may be another matter—that is something you should evaluate during your service trace. Bad enough for your public image if your web site is inaccessible due to a cyber-attack, but if you have gone to IP phones and no one can reach you by phone, either ... you will be chaff, and probably unrecoverably so in people's minds.

Data Currency

One final aspect to be aware of: When you activate your network continuity plan, the alternate location, whether on-call or internal, will represent your company to people who formerly were dealing with the now unavailable location. Will both parties to the conversation be operating from the same set of data? How often are your databases replicated? Lehman Brothers obviously needed constant replication, but that is expected in the financial markets. You must know how current your data needs to be, then check to see if that level of currency would be available.

If your industry's norm is to enter changes to a file which will be posted to a master change file during a maintenance window every night, and the resulting summary update is posted to all applicable servers, you must consider what changes may have been entered before the network location was lost (whether to physical reasons or a cyber-quarantine does not matter). If those changes were not written to the summary file used to perform the update, customers, partners, and vendors will not understand why they had a perfectly valid transaction—here's the case number, they might say—and you have no record of it. You should have in your plan how such situations will be handled; you may choose to simply take the customer's word for it up to a certain valuation, require a supervisor's approval to a higher level, and so forth. You may provide a script which the lucky employee may use to explain the problem and the fact that there will be a delay of one more business day in the transaction.

Trade-offs

As you develop this plan, you will come to areas where better would be nice if it were only affordable. You may not be able to redesign your network to provide internal redundancy for another five years (although you can migrate to that over those five years, a topic for Chapter 10, "Returning from the Wilderness"). In that case, even though internal redundancy would be better, you will make do with an on-call leasing arrangement because that is what you can afford.

Affordability and costing alternatives will be a subject for discussion in Chapter 11, "The Business Case." The better you have documented

- What the network does
- What it must do
- What it currently can do
- What it could do if these actions are taken

the better your case to get your preferred choice. That doesn't guarantee you will get it, of course; nothing does. But why not maximize your chances?

CHAPTER 10

Returning from the Wilderness

The Promised Land always lies on the other side of a Wilderness.
Havelock Ellis

If your network continuity plan covers only how to act when disaster happens, it is incomplete. Life goes on; what is open to question is your degree of participation in that. What will be your participation the day after disaster, the month after disaster, and the year after disaster? The details of the effects with which you must live will differ in the event of a cyber-disaster compared to a physical disaster, so we will consider them separately.

Even though we consider them separately, they do have common threads running through them. In both cases, there is an immediate problem of network operation in a suddenly degraded condition. With a cyber-attack, you must be concerned with forensic procedures, if for no other reason than to keep such an event from happening again. With a physical loss, that is not an issue, but sustaining operations after the first surge is; we can all pull together and work extra for a few days, but when it becomes weeks or even months, that is less tenable.

After we understand what hit us and how we will deal with it over the next few months, whether we lost network usage from a cyber-event or a physical event, we must, sooner or later, restore our operations to at least what they were before the disaster. And, finally, this is a powerful opportunity to evolve the network to what we want it to be, sooner than we might otherwise be able to get there ... if we are prepared to think in those terms.

Cyber-Recovery

Once you realize you have been attacked, two efforts must proceed simultaneously. They are not mutually exclusive, but, without forethought, they can get in each other's way. One effort is operational, while the other is forensic (whether or not you ever choose to take any legal action).

Operational Procedures

Your primary operational tasks are to determine where you are compromised and the nature of the compromise (in as much detail as possible) and to undo that damage. All three of these are, of course, far easier said than done. Systems logging is important, but so is the old-fashioned, pen-and-paper kind by the people tracking down the parameters of the problem. Such note taking prevents duplicative effort, especially as people become tired and frustrated; should you choose to pursue legal action, such personal logs provide a far better foundation for testimony than personal memory, especially considering the amount of time that will probably elapse between the event and the litigation. To have that option later, you must make the effort in real time.

You have a cyber-attack plan, which includes activating a response team. They have access to the resources that they need both in terms of the information and the hardware and software. If what you have detected is a form of brute-force attack (such as Nimda, Code Red, Li0n, or other rapidly-propagating worms), your personnel will rapidly sort out into groups. One group will be frantically trying to determine the extent to which the worm has already reached so that the logical area can be contained (a cyber-quarantine); to preclude reinfection, another group will be attempting to purge the worm/virus from infected hosts inside the quarantined zone. Yet another (if you have the luxury of a third group) will be attempting to ensure that hosts in the path of infection are vaccinated—their anti-virus software has the latest updates, OS and application software have the latest patches, and unnecessary services have been removed. The firms that provide anti-virus software are generally quite prompt with a profile update, but it does take time to download and to install it where needed, as do patches. This is an argument for a separate (and separately protected) management network or VLAN, to be sure that you can get out to the anti-virus site and then distribute the update among your critical network elements; it is also an argument for staying abreast of the manufacturer's updates for the software you have loaded.

If you have the staffing for only two groups, have a quarantine team and a vaccination team; treatment can wait longer than the containment and prevention of spread. This means that when upper management calls because they need their access restored, you must do three things: buffer them from your team, explain that they will have access more reliably if the two higher-priority tasks are completed first, and not remind them that you asked for more staff and better software in the budget requests and were turned down. Despite the poetic justice, not only is now not the time, but if you defer that and make it a part of a reasoned and well-crafted post-mortem, it will carry more weight.

If the attack is more subtle, and you realize files have been tampered with, backdoors are planted, proprietary information has been accessed (and presumably copied), your problem is similar in concept. The pace will differ, though, as there is unlikely to be the pressured atmosphere resulting from loss of access by many users, a large portion of whom are calling to notify you of their outage and then to wonder when they will be back online. Another difference is likely to be the location of the damage (everywhere versus a few select hosts—which must be tracked down) and the sophistication of the modifications made to your network.

Containment in either case buys you the time in which to begin the recovery. For this effort, you can proceed based on what you found as you worked through the operational problem. You will probably do a better job, however, if you have carefully collected the forensic information along the way.

Forensic Procedures

In any particular attack, it is not a given that you will take legal action; what is given is that, should you subsequently desire to take such action, careful collection and preservation of certain information about the event and how you worked your way through it will make or break your case. Because you have only the one opportunity to collect such information—you cannot go back later and recreate the data with any fidelity—you must include in your cyber-defense effort your procedures for such collection, and you must practice them. They do not come naturally to most people. You may not necessarily know what information you should collect, from a legal standpoint, though you are aware of what you need for operational forensics. Consult with your legal counsel; you will probably find that the operational data you should collect is the same, but how you collect it and/or how you can later prove how you collected it is just as important legally.

Likewise, it may be advisable from a legal standpoint not to disclose to many people all of the information you collect. It may be useful to keep some of it extremely close-hold, lest (for instance) you give away key information that could decisively identify the culprit (along the lines of the World War II saying, "loose lips sink ships"—the information was not always classified, but it had great intelligence value nonetheless). An example of this would be if you identify some utilities left by the hacker, which are known to be associated with a particular scripting tool. If the hacker hears through the grapevine that you know these things, he or she may return and remove other items that would be damning—and you will therefore never find them.

Personal logbooks are a very useful tool. In addition to aiding the operational effort by validating that something has or has not already been tried, they establish timelines. Entries must include an initial date and time, and every subsequent entry must include them as well. All entries must be nonerasable—use ink, not pencils. You must have a standard of how time is reported, whether it is local time zone, HQ time zone, or UTC, and a standard time reference. System logs should also report on the same time zone and according to a standard reference. Standardized, consistent, and reliable time references will enable the chain of events to be described believably.

As you counter the problem, though tempting, try to avoid using a whiteboard. A paper chart, with the pages still attached to their pad (folded over rather than torn off; if you must tear them off, annotate them to establish their order) also establishes the sequence of idea development as you traced the intrusion through your network. Paper also preserves sketches instead of erasing them to make the next one. Another type of permanent record, rather than an erasable one, is data captures. These should be on read-only media, if possible (write once, read many times like a CD-ROM archive format).

Above all, you are trying to be sure that 1, 6, or 12 months from the event, you can go back and reliably reconstruct what happened, when it happened, and where it happened. Taking legal action will be problematic for several reasons. Proving real dollar-value damage to a jury from the general public, in a trial presided over by a judge who has trouble with email attachments, in the face of an opposing attorney who only needs to create doubt, is a very difficult proposition. Complicating matters is that careful attackers (the ones you most want to prosecute and those who, from the network administrator's perspective, most need to be punished) often route their probes and attacks through several intermediary systems. It is difficult enough if those several jurisdictions are all in the United States, but it is likely that they will cross international boundaries.

The odds are against your being able to do much legally unless you have airtight documentation and demonstrable, serious damage that the most senior leadership of the company is willing to state publicly. Even then, it will be a long and arduous path to legal justice.

Nonetheless, the forensic approach to data gathering during the event is critical: It is how you will analyze after the fact what happened and how it happened. Unless you want to go through this sort of exercise both early and often, you must conduct a post-mortem so that you may not only correct your mistakes, but correct the right ones (be effective as well as efficient).

To conduct the post-mortem, you will want answers to the following questions:

- How you were penetrated
- When you were penetrated
- The extent of the compromised systems
  - Which systems
  - How they were compromised
  - Altered utilities
  - New files (Trojans, back doors, sniffers, password collections, and so on)
- Who compromised them
- Why

Note that, in order to (reasonably) safely return the system to operation, you need the first three major points but not the last two. If time is a problem, if for business reasons the network must be back in operation by an undesirable (to you) deadline, knowing the first three tells you how to keep the intruder or his or her many friends from coming back through the same unlocked door, it tells you which backups are probably safe to use, and it tells you which systems must be repaired and the extent of the repairs. As gratifying as it might be to know, the identity of your attacker and his or her motivations may remain a mystery, even if time is not a factor—some attacks are simply not definitively traceable. Without knowing who, knowing why can be inferred only if certain files were targeted. And, like the U.S. Navy's ambush of Admiral Yamamoto in World War II, a skillful attacker will make it look like those were not the target, but a lucky break while doing other things.

To collect the information you will need, you must organize your people along lines they may not be used to on a daily basis. And you must do this while the world is coming down around your ears; planning would help, so start that now. You will need a person (two for each task would be better, but realism suggests you'll do well to have one each) to do the following:

- Continue monitoring the network—the intruder may still be present (and watching) or may return.
- Perform backup validation/testing/copying (some attack software includes a routine to erase the backup when it is mounted).
- Identify any compromised systems.
- Prioritize system fix/cleanup—your service threads will help here (the DNS server before the mail server, for instance—or how will the mail server resolve addresses?).
- Establish the time of the attack.
- Establish the method of the attack.
- Identify and document any modifications left by the attacker (other back doors, compromised files/services/utilities).

While your team works on all of this, your job is to do none of it, no matter how much (for instance) they need an extra pair of eyes going over the logs for information about the attack. For one thing, as the manager, you will be constantly interrupted, and crucial portions of the logs could be missed on the assumption that you looked them over when, in fact, you skipped that part inadvertently when you were distracted. For another, you have a more important job.

You must be a buffer. With the network down, people will suddenly discover how much they miss it. While the help desk can fob off the routine calls, they cannot do that to the calls from senior management; you must. Of course, you cannot really fob them off when they call, but you can prevent the calls from taking excessive amounts of your time by keeping upper management advised, on a periodic basis, of the status of the cleanup. Again, you may not want to keep them all informed about the detailed status of the investigation; you should obtain legal guidance on that, appropriate to your business and jurisdiction.

You must also know in advance what will constitute victory conditions for this effort. In war gaming (commercial as well as military), there are certain criteria that, once met, define the end of the game because it has become a given that one player will win. While the term victory may seem less than appropriate here, you must establish a similar set of criteria to define the end of the forensic effort. As one security consultant noted, one hour after the network crashes, senior management will be conservative and want you to get it absolutely right. After a month, however, they want the network back up, yesterday. Establish in advance the minimum criteria that you must feel confident about before you will put production traffic back on the network or reintegrate this site or subnet with the rest of the network.

These criteria will constitute what you deem an acceptable level of risk for the network's operation. Short of rebuilding every host from scratch and recreating (not restoring from backup) all data files, you cannot be certain that your network is clean. Even that, of course, assumes that no insider within your organization was responsible for this problem or decides to take advantage of the rebuild to make things easy for him- or herself to hack the network later. It would be easy for you to become paranoid; to avoid this you must decide in advance how much risk you are willing to tolerate in the network.

Risk tolerance varies from one individual to another, but in general, it follows the function in Figure 10.1. Initial levels of risk do not require much reward (the risk of failing to receive the reward is low, and the reward itself is low, meaning that it would not be missed much). As risk increases, it requires more and more potential reward to be tolerable. As a logarithmic function, we know that ever-increasing levels of potential reward are required to assume lesser incremental risk. At some point, the rate of required reward increase so greatly exceeds the rate of risk increase that no further risk will be taken on. That point varies among individuals, among companies, and among entire industries.

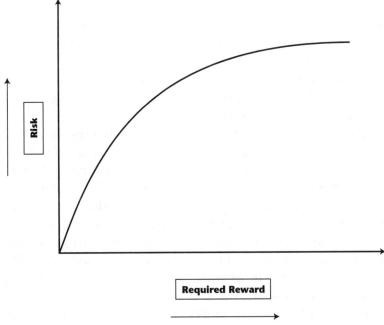

Figure 10.1 Risk tolerance.

The risk tolerance should not be based on your opinion alone. Consultation with your senior architects should obviously be a part of developing your criteria. They should also be based on clear agreement from senior management as to the overall degree of risk the company is willing to bear. Within that overall risk, you will have to make trade-offs among the nice-to-haves (necessities you can't trade away; if you make it to the luxury category, you're more fortunate than any manager I've ever known).

With this kind of advance planning, you can think more clearly during the attack about what you must do. You may even have the luxury, like the Managing Editor of *The Wall Street Journal*, of thinking about what you will deliver tomorrow.

Physical Recovery

To a certain extent, the high-level problems of physical recovery are not unlike the problems of cyber-recovery: What do you have, when will you have more, how reliable will it be? The details, though, differ considerably.

Immediate Operations

In the short term, right after the disaster, you have two very different priorities on your mind—your people and your network. You know that, human decency concerns aside, you need the knowledge and skills so painstakingly acquired and resident in your people. They are the asset most difficult to replace; if your routers and servers are trashed, the equipment can be replaced in a few weeks, with data and configurations reloaded from backups and saved documentation. People have a far longer lead time to be competent, especially in a field as fluid and demanding as IS.

At the same time, you must be worrying about your network—how well the redundancy is working in practice as compared to your plan. If the disaster occurred at your location, you won't get to know the answer to that for some time; how much time depends on the nature of the disaster and what has happened to communications in the process. If the disaster is at one of your other sites, you must resist the temptation to help the recovery/network reconstitution. Again, you will be the buffer between your people and the questions from above. Further, someone has to be thinking while others react, and someone must work within the plan as much as it fits the actual event. Even those with well-considered plans had to adapt after the World Trade Center disaster; you need to be leading the adaptation team, not helping the hands-on team.

The immediate task is to activate the network continuity plan, if it hasn't been activated by someone else. This will depend on the criteria you have set for activation: authority and the succession of that authority when the designated person is not available. Once the plan is activated, you or someone in a location to do so (because yours may be the one out of operation) must monitor the process, ensure that communications come up, and interested parties are notified. Some tweaking must inevitably occur in real time; no plan is an exact fit for how events actually unfold.

More important is the transition from the immediate aftermath into a sustainable operation. That is where you should be looking as soon as the plan is activated.

Sustained Operations

Especially if you have elected to use a leasing arrangement from an organization such as SunGard (Computer Computer, Ltd. operates chiefly in the United Kingdom but can provide a similar service on the continent and in North America; corporate system vendors are entering the market, as well), you must transition within a few weeks from the temporary location to a new one. That means acquiring access to real estate, replacing hardware, getting your software loads on it (including configurations, especially security-conscious configurations—some hackers deliberately target companies that are likely to be in disarray), and establishing communication paths from the new location to the rest of your network.

If you have chosen the internal redundancy approach, as *The Wall Street Journal* and Lehman Brothers did, you probably have a little larger window in which to replace the lost location, though not a lot larger. The other sites that are carrying the load were not necessarily designed for that, and they should not reasonably be expected to operate at that level for a sustained period of time. At the same time, you have now lost a significant portion of your internal redundancy; until you replace that, you are vulnerable to a further loss (physical loss or, as noted previously, loss from a cyber-attack, whether opportunistic or a less-than-ethical competitor).

Therefore, a part of your continuity plan is to include what happens after the immediate need for continuity has been met. You might want to think of it like a play or a movie: Stories have beginnings, middles, and ends. Successful stories have twists at each transition that take the story in a new and unexpected direction. Successful stories are also like life—they pass our reality checks (or, at least, don't require us to suspend disbelief too far). It should not surprise us then when real life transitions from one stage to another with a twist; of course, it does surprise us because we tend to assume predictability, even when we theoretically know better.

As you make the transition from immediate operations to sustained operations and, ultimately, to restored operations, expect twists that cause you to rethink or rearrange part of your plan. You may have included a follow-on location to use after temporarily working out of a leased service facility, but the disaster that disrupted your principal location has rendered new communications links to the follow-on site impossible to obtain in a reasonable timeframe. Or perhaps some or all of the major transportation arteries connecting to that area are down; getting there would be more than half the fun for your people. Perhaps that facility is also a casualty of the disaster. Office space in Manhattan, not in overwhelming supply before September 11, 2001, became extraordinarily difficult to obtain after the disaster: So many firms had to relocate, not just from the World Trade Center but also from buildings in the extended vicinity, due to damage, ongoing rescue and firefighting operations, and transportation disruptions, that existing supply was essentially consumed in very short order. If your intended new location had been two blocks away from the World Trade Center, you would have not have been able to get into it.

With a little luck, however, you can actually turn the situation to your advantage if you can take the time to organize what you are going to do as you move from immediate to sustained to restored operations.

Restoration

Restoration to a full, and fully operational, network is your minimum long-term goal. We hope that it will be a graceful reversal of the changes made when disaster struck. In the spirit of making lemonade when life hands you lemons, you would really like to take this opportunity to move the network along an intended, evolutionary path. If you have totally lost a location and all its facilities and equipment, then when replacing the lost functionality, you may be able to create an upgrade that you would not have had, certainly not so extensively. You would not be the first to draw such lessons.

The Wall Street Journal took the lessons from the New York City power failures to heart and not only restored normal operations for its existing network design, it redesigned to ensure that it would never again be vulnerable to the loss of a single location. Merrill Lynch had run a planning exercise concerning hurricane preparedness less than a year before the World Trade Center disaster; lessons learned from that natural disaster plan enabled it to have 90 percent of its New York staff relocated within one week after September 11.

Restoration then can be used to return to the *status quo ante bellum*, or it can be deliberately used to move the network forward. Which you choose will depend on whether you have a planned evolution of your network already available that you can adapt when operating under the stress of sustaining operations in a non-normal location or with a non-normal structure. Remember that these stresses do accumulate, and sustained stress affects people's judgment and teamwork. Including yours—perhaps especially yours, if you truly feel the responsibility for your people and your operation.

Before you begin restoration, there are some things you need to examine. Whether you have chosen to take legal action against whoever caused you to lose a portion of your network (assuming you can reliably and *provably* identify said person or persons and that legal action has a reasonable probability of leading to recompense), you need to conduct a post-mortem. It should occur enough later that you can gather the information you need—it cannot be based on guesses and suppositions, though some of those will likely be unavoidable—but there should not be such a long time lag that details get lost in the press of subsequent events.

My rule of thumb is a week after things are stable, more or less. The week gives time to clean things up a bit from the initial disorganization, which will occur, only varying in inverse proportion to your level of preparation, and it gives you time to have people thinking as well as looking at things, gathering the information. The things you want them to think about are these:

- How close was the event to what we envisioned when we made our plan?
- How well did the plan work?
- How well did we implement it?
- What did we not expect that we should have?
- Where did we get lucky?
- What could have happened that would have thrown everything askew?
- What did you do as management that helped? And what hurt?

Though some of these questions will seem similar, in fact, they ask subtly different things. How well the plan worked will be assessed in conjunction with how well it was implemented. The latter, of course, subsumes faithfulness to the plan, and both effectiveness and efficiency in carrying it out. It is important to evaluate the plan separately from its execution—assume that it had been executed flawlessly: In that event, would it have actually been a good plan?

The fourth question seems self-explanatory, but this is a chance to examine the assumptions that support your plan. If you should have expected something, knowing what it was is important, but understanding why you didn't expect it is far more important. Assumptions, like first principles, drive everything that follows. Make the effort to consider what you took to be a given; you may have been right and you may not, but you won't know the answer to that without asking the question.

Like the first question pair, where you got lucky is not quite the same as where things could have gone wrong. The former asks about things that actually did happen, though you were not harmed by them (or not as much as you might have been); these are statements of fact. The latter asks what else might have happened but didn't; these are matters of conjecture. When you answer the second question of the pair, include what you likely would have done in each instance based on the information known or guessed at the time when the event would probably have occurred and whether, with hindsight, that would have been a good decision to make.

The final question is really another pair, and it will be answered honestly only if you have already created a culture that knows it is safe to tell the truth. This is your chance to learn unpleasant truths, certainly, but some of the answers (to both questions) will surprise you. And the truths may improve what happens the next time.

The results of the post-mortem must be reported to senior management, though typically you will not need to go into great detail; however, you must identify the critical issues. This report is where, with facts in hand, the results of fiscal inattention can be laid out without whining. This is also where you may use the plan for restoration to move the network forward rather than merely returning it to where it was. How you present that will depend on your knowledge of the company's fiscal health relative to your needs. Again, you will have to sort carefully to present necessities to regain what you had before the disaster, nice-to-haves (marginal improvements), and fundamental improvements. Be prepared to defend what you call necessities; risk analysis, even on a qualitative level, will help you make your points.

Using restoration to improve the network will be a more difficult sell if the event was a cyber-attack. Nonetheless, the improvements are likely to be more procedural and involve the implementation of existing software rather than new purchases. That makes them easier to sell, especially if the initial and ongoing cost is compared to the cost of lost network function. We will spend some time in Chapter 11, "The Business Case," on how to cost these improvements and a proof-of-concept exercise to demonstrate the cost of network nonavailability.

With that in mind, restoration may take only a few weeks in the event of a cyber-attack (or even a few days, depending on the extent of the rebuilding you need to do). It may take months if you have had a major physical location destroyed; don't forget to include notification of customers, vendors, and partners about your new location and phone numbers. While you may have a preferred-customer status with all your hardware vendors, you will need another physical facility of equivalent supporting capacity—square footage, power supplies (note the plural), communications connectivities (plural again), and more. Even arranging the lease could take a few weeks; getting your utilities connected may take more time, especially if the disaster was widespread. Only then can the equipment be delivered, have its software loaded and configured, internal connectivity established (running cable is *always* popular among the staff); external connectivities must be installed and tested—do the backup/failover procedures work? Does the UPS kick in and the generator come on when line power fails? Do the circuits actually go where the carrier said they would?

And so forth; the configuration and testing phase for a new facility (new to you, even if the building itself is not new) will seem excessive, and you must be able to document all the work that is going into it. That means your network continuity plan's restoration phase should lay that out in advance. After a disaster is not a good time to surprise senior management with how long it will take to make use of the new facility.

Undress Rehearsal

Exercises will prepare you and enable you to test your plan, at least in principle; you really can learn lessons from disaster while experiencing it vicariously. You must make the experience as close to the real thing as possible without incurring real damage. This is the fine line trod every day by professional athletic teams; it can be done. And, like sports or drama, many practices need not be full effect; they are intended to get the feel, the timing, the coordination and teamwork down pat. Sports teams do walk-throughs or drills at less than full speed, in less than full game kit. Actors rehearse pieces, get them down, then pull together pieces into scenes, and finally scenes into a full performance. But they practice that at less than full kit—an undress rehearsal—before they expend the resources and intensity of a full dress rehearsal. You will need to borrow a page from those performers—athletic and thespian—when you practice your plan. You will probably never get to have a dress rehearsal—taking down the network to see if you can recover it is considered bad form.

Make good use of your undress rehearsals.

The exercises you will probably be able to conduct will be small in comparison to an actual disaster and involve only a few people. These practice sessions are overhead after all—an expenditure of resources that does not directly contribute to revenue. That such practices may prevent the loss of direct contribution to revenue—and a large loss of contribution—is your justification.

Your exercises will necessarily be time compressed, as well. Expect to go through the entire cycle of a disaster or a network attack in half of one work day, no more. Until you have enough practice under your belt to be sure you have the teamwork and knowledge immediately available, I would recommend a biweekly exercise. Assign problem development for each exercise to a different pair of people, and they will serve as the entire outside world for the benefit of everyone else during the exercise.

LOST CONTRIBUTIONS

As an example of the lost revenue contributions, a network supporting financial services may well support activity worth $100,000 to $1 million per hour. During the Code Red worm (July 2001), one nonfinancial firm with a large presence near where I live lost most server access throughout its global network for 1.5 days (not all servers had proved vulnerable, but the only way to prevent propagation from unknown servers people had set up on their own—just trying to get the job done—and quarantine the affected areas was to physically isolate large subnets).

Assume 20,000 employees with their productivity halved for that long (the firm has largely gone paperless, in part due to its global nature, and files are routinely exchanged worldwide). Two weeks later, Nimda struck, exploiting holes created by Code Red. Another lost day.

Incidents like these lead to individuals keeping more of their work on their local hard drives, rather than on the network, just in case—which contributes to data loss in the event of a disaster as well as the convenient workaround of simply sharing out their files when large ones need to be exchanged. Shared files leaves a Windows host open to attack if a hacker gets into the network; recall Chapter 4, "Murphy's Revenge," and the development of a DDoS attack.

Your exercises may lead you to weaknesses in the network before an attacker exploits them. Loss prevention is hard to quantify, but there is some help for that in Chapter 11.

Start small, perhaps with a minor hacking scenario—the ideas behind that should be reasonably familiar. Encourage your exercise team to throw in twists. Even when the twists seem unrealistic, go with them on the assumption that you don't necessarily think like a hacker; he or she may have some motivation that just didn't occur to anyone in a white hat. You may wish to alternate the scenario types—a network attack exercise will be followed two weeks later by one simulating the total loss of a major site. Different skills will be developed for each type of exercise, but there will be spillover effects.

Who gets to play? Like nearly every other question in networking, the answer to this one is "it depends." I generally recommend getting the managers up to speed first, for at least three reasons. First, they are less likely to be missed for a few hours, while line administrators may have production work that should be done. Second, as these people get an idea of what skills they need to have in their personnel when things go bad, they can start developing those skills in their people. You'll thus have a fairly quick beneficial effect on personnel readiness to handle glitches in the network. The third reason is that you will have, as a rule, people ready to work hard to handle a situation, if they only knew what to do. Getting the team leaders/managers organized and working together makes teamwork in actual execution more likely; there ought to a fairly rapid gain in efficiency compared to choosing a few line operators to train first.

Unless you have someone with prior military experience or a dedicated gamer, your early exercises will not be as well structured and planned as you would like. Without practice, though, your plan is largely a wasted effort; exercises not only prepare you and your leaders, they test the plan. To that end, keep working on the exercises as a means of improving both. Here are some tips for creating exercises:

- Think small (at least at first).
  - Exercise/test a piece of the plan, not the whole thing.
  - The exercise director will need some time to prepare a scenario.
  - Scenarios should have some basis in reality.
  - Scenarios should include likely responses to your requests for information or help.
  - Exercises should include one or two twists (these take practice to make believable).
  - Information comes to the team being exercised only when they earn it.

- Order data collection.
- Ask for information from outside sources.
- Timelines will be compressed.
  - This requires information from the exercise director to the team as to elapsed time/what the now-simulated time has become.
  - Expect to simulate approximately two days.
  - Expect to test personnel management plans as well as technical management.
    - Remember, generals study logistics—food and time off will be major factors for the people trying to restore your network, both in effectiveness and efficiency.
- Include in the exercise planning some time for a debriefing/evaluation.
  - This is not a forum for the exercise director to display his or her cleverness in stumping the team.
  - The goal is the same as a real post-mortem.
    - What did we do right?
    - What did we not do right?
    - What does that mean with regard to the plan (versus our execution of it)?
  - Pay special attention to why things fell out the way they did (guess versus knowledge, for instance).

Here are two sample exercises, one cyber and one physical.

Exercise Scenario 1: Cyber-Problems

The simulated time is 0418. Your team has been activated because a network manager from STL has called, very angry, wanting to know what the devil (language edited) you're trying to do to his network. Your midnight shift admin managed to get the following information: STL has just lost its last two attempts to carry backups for paying customers when its network has been flooded by malformed packets. The packets are coming from about 1,500 hosts, more than 400 of which are in your address block, according to ARIN (American Registry for Internet Numbers; URL in Appendix A, "References," under *Miscellaneous*). If it happens again, STL will charge you the penalties it will have to pay its customers.

Your team immediately assumes that you have been penetrated and that 400+ hosts have become zombies for a DDoS attacker.

This exercise can go in either of two directions at this point:

1. You actually have been penetrated and have a large number of compromised hosts; guilty as charged.

2. Someone else's hosts are the zombies, and a twisty part of the problem is that the zombies are spoofing your addresses in the source address field; innocent of all charges.

If you take the first direction, the problem is to find the zombies (perhaps STL's administrator was too busy as well as too angry to provide a convenient list; you may be able to get some representative address blocks if the exercise director is willing to let you be lucky). Next you must quarantine them, disinfect them, and find out what else has been done to your network, as well as finding the opening that let in the hacker. Another possible twist is that there is no opening: The hacker is one of your own, one who just brought the problem into work on a floppy.

If you choose the second direction, the fun part is proving your innocence. System logs are an obvious first choice; the logs in question would be from your egress routers. A twist could be added that logging has not been turned on in two of the routers (they may be misconfigured, or logging was not a part of the configuration file that has been loaded). Without that, how do you prove your innocence? Perhaps you have a good relationship with your ISPs (plural; you are multihomed, I hope), and they have logs of your outgoing traffic in the relevant time period.

The second choice permits a shorter compressed time line than the first. You could also take the second path one time (perhaps the first time, to prepare everyone to think of the nonobvious as well as the obvious); two or three months later, when you are better at this sort of thing, run the same exercise, but use the first path instead. And, because you are collectively better, the exercise director can have the hacker leave some unpleasant surprises behind.

Exercise Scenario 2: Physical Problems

Corporate headquarters recently moved into a very attractive business campus along the Pipsqueak River in Semiquaver, Pennsylvania. A little downstream the river valley narrows between bluffs, but the business park sits on a lovely flat bench in a wider part of the valley. At least, it's lovely in the summer. This winter has been ugly, with large parts of the river actually frozen over; everyone is grateful for the signs of spring. Snow has been thawing rapidly, and that will only accelerate with the warm rains this weekend.

You are at a St. Patrick's Day party (for some people, any excuse will do for a party) Sunday evening when your beeper goes off. You call the duty network administrator, who is nearly panicking. The police have given him 15 minutes to get everyone out of the building and to higher ground. An ice dam has formed on the Pipsqueak, only a half-mile (800 meters) downstream, and flooding will surely follow.

Your team should know that you run a series of backups to a remote facility on Sundays, from 1600 to 0000 (midnight). The backups will be half done, at best, and you are about to lose all access to a major network facility.

Factors to bear in mind for such an exercise include these:

- How much equipment is left on all night? Can the staff on site get it all turned off?
- Do you have batteries for backup, which might explode when flooded if a spark is handy?
  - Will electricity be cut off first? You should hope so—firefighting in a flooded plain is nearly impossible, and you might lose the entire building.
- How long will you have no access to your building?
- Where is the most valuable equipment, in terms of elevation above the flood plain?

This is a rapid evacuation exercise, with probable extended loss of use of the facility. Fortunately, there are few people to evacuate (and account for), but many to notify not to come to work, as well as the notification to senior management of the disaster, of course. Depending on the degree of flooding and if there is a fire (there were fires when Fargo, North Dakota, flooded from an ice jam), you may keep the building but lose all the valuable electronic gear, as well as paper files, furniture, and more. Time compression for this exercise will probably approximate several days, even weeks.

Again, this is a fairly simple scenario, enough to get you started on physical problems. You could posit many other scenarios, appropriate to your physical locales and their political climates.

Evolution

Your business has a plan of some sort for its growth and evolution over the next few years. You should also have a plan for the evolution of the network that supports that. You can use your network continuity plan to further the business's continuity plan by using the network continuity plan to evolve the network to better support business growth.

This may sound convoluted, even Machiavellian, but it is sound business practice to get the most value for your expenditures, and that's what this is about. We can use CDG as an example. Recall its logical topology, shown here in Figure 10.2.

We also remember that CDG has centralized all DNS service in the Kansas City NOC. Network continuity is provided by rolling services (except DNS) clockwise; backup intersite frame-relay connections are available with notice. Physically separate circuits connect each CDG site with its local Secured Storage facility. CDG has growth plans, assuming its contract with the home improvement chain is renewed after the six-month trial; its target cities are Memphis and Fort Worth for the first expansion.

CDG has acted as an intermediary between the carriers and end customers. In part, that's because customers were fed up with the service and responsiveness of the carriers. As a larger account, CDG obtained better service. But the market seems to be flattening: Carriers are trying to move into the role CDG has occupied. The choice is to be even better, and thus keep the carriers at bay, or acknowledge defeat and exit the business as gracefully (and profitably) as possible.

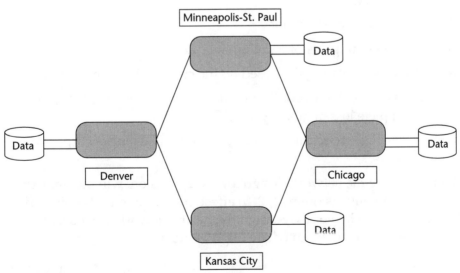

Figure 10.2 CDG logical topology.

CDG has decided to fight, and its weapon will be quality and the ability to deliver faster and better than the carriers; Shelly Guo and her partners believe that, as long as CDG's prices are not too different from those of the carriers, service will be a sufficient differentiator. As a result of some exercising of the network continuity plan, Shelly realizes they need to improve redundancy in a few key areas: power supply, DNS, and alternate connectivity for customers if a CDG NOC goes offline. CDG's expansion will require the assumption of debt, which makes nonperformance penalties (even if they are credits on customer accounts) untenable. Credits on customer accounts mean a reduction in revenue, and Shelly and her partners became very well aware early in their business venture that cash flow is critical. They will not be able to service the debt unless they can rely on their revenue.

Shelly's job thus becomes to improve network reliability as inexpensively as possible (to conserve current cash flow) while also improving the network continuity plan. With those improvements in hand, CDG ought to be able to expand without endangering cash flow. Shelly has three months to get this done so that they can demonstrate their continuing improvements to the important new customers, thereby increasing their chance of contract renewals. Her improvements—which should help make expansion possible—must also fit with a resilient, reliable expanded network, including Memphis and Fort Worth; CDG does not want to be inefficient in its growth and make more work (and less profit) for itself. Her tasks, in order of immediacy, are as follows:

1. Improve DNS.
2. Improve power redundancy.
3. Improve existing customer alternate routing.
4. Evaluate security (again) based on the changes in the network.
5. Improve the network continuity plan to reflect the possibility of the complete loss of one city's NOC.
6. Rigorously exercise the network continuity plan and incorporate the results.
7. Reevaluate the network design for the expansion; will the immediate improvements (items 1, 2, 3) affect migration to that design? If yes, decide whether to redesign the future network or reconsider how to execute the immediate improvements.

Notice the gotcha in task #7: She must plan her tasks in advance and realize if she has interdependencies. If she executes on tasks #1, 2, and 3 only to find they make the planned expansion more difficult (perhaps due to contracts for customer alternate routing becoming inefficient, given the new design), she is now forced to redesign the expansion or re-execute the immediate tasks. Both are a waste of resources.

Take the time to consider ripple effects, and make the improvement plan and the continuity plan an iterative process involving future network design plans as well. As an example of this, we can look at the plan to improve DNS (see Figure 10.3). Because Denver is the corporate backup, Shelly will simply accelerate making it the secondary DNS authoritative server. She plans to have the senior network administrator in Denver obtain a book on DNS (Kansas City's DNS administrator strongly recommends one he calls "the Bible for DNS"). In two weeks, the Kansas City DNS administrator will fly to Denver to help set up (only help, not do it himself) the new secondary server. When it is complete, the secondary server in Kansas City will be demoted to an ordinary DNS server. In the meantime, in case something should happen to the (one) DNS administrator, she directs him to immediately train an alternate. To assure him that his job is not being threatened, Shelly points out that he will have someone with whom to share the pager for DNS problems.

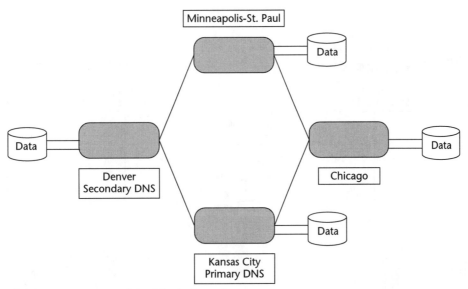

Figure 10.3 Improved CDG logical topology.

The person selected in Kansas City has a dark and devious mind; she's been the one responsible for the best continuity exercise scenarios. She advises Shelly that she intends to plan one that has Denver's DNS server on a maintenance window (and off the air) when Kansas City goes down. That takes CDG back to the original problem: no available DNS server. CDG has not been able to pay as well as some competitors, and Shelly has been reluctant to pass much high-level training on to the administrators in Minneapolis and Chicago because their turnover has reflected CDG's relatively low pay.

Shelly faces the fact that she must let go of some of her close control over network operations by devolving more functions to all sites. A DNS server will need to be available at every location once the company expands (see Figure 10.4), so they may as well make that change beginning now.

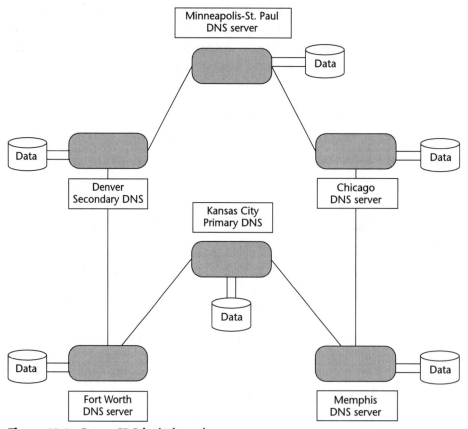

Figure 10.4 Future CDG logical topology.

Training and investing in people is expensive, especially when the investment leaves for a better-paying job. To bring pay of their personnel up to market levels, CDG must increase its senior administrators' pay by 8 percent; without the pay increase, the training expenditure she's about to make will be wasted. To find the funds to make the pay increase, she has to reevaluate how she is going to improve redundancy and reliability.

After running the figures for several possibilities, she concludes the following:

- Redundant connections from the customer are really the customer's responsibility—they must either choose to be multihomed or bear the risk of single-homing themselves.
- CDG can protect from the loss of an entire site by contracting for on-call facility support from a provider like SunGard.
  - The contract will provide supportability for one site but will apply to all CDG locations.
  - Configuration information for all CDG equipment for all sites will be kept in soft copy at every site, with strong access control; when a backup location comes online, its senior administrator can access these files at any remaining CDG location and make the required connections.
- All sites will provide all CDG services, at a consistent quality level, reducing the impact of a single lost site on the network.
  - The increased training required will be offset by more productive use of management time in the main office.
  - Shelly will no longer need to hire another administrator in Kansas City because she will not have to pass some of her oversight responsibility to her most senior administrator there in order to get the rest of her work done.
  - Oversight will be done on a more site-centric basis, with weekly meetings via conference call and shared desktops (CDG uses NetMeeting for this).
- Power redundancy must be improved to true UPS at every site, with a backup generator (switchover tested monthly in a maintenance window).
- With more site independence, security must become site-centric as well, with the same (reduced) level of oversight from Kansas City.
  - Network security procedures will be exercised monthly at each site.

- Exercises will occur on a rotational basis.
- They can be more extensive as an adjacent site can take some of the customer traffic load.
- They can be evaluated more fully and the lessons learned passed to all other sites—which will be exercised to see if they incorporate the lessons learned.

CDG can grow—and survive in the face of increasing competition from its own partners in service provision (the carriers)—only if it does two interrelated things:

- Devolve more authority to subordinate units so that headquarters can manage the whole company instead of micromanaging pieces of it.
- Evolve to a more distributed architecture that provides a great deal of resiliency to network operations; where the resiliency is not sufficient, as in the case of a loss of use of one complete site, backup service is contracted for.

CDG has reached a critical juncture in its business life. As when a person first becomes a manager and learns the hard way that he or she cannot do everything himself or herself, CDG must let go of some centralized control and delegate. The delegation process, necessary to business survival, can be improved by using network continuity planning to evolve the network to a more distributed system, where each site is interdependent with the others, but with far fewer total dependencies.

The network will be more resilient in the face of disaster. As a result, it is more likely to survive the unexpected. And the business is more likely to still be in the game when life itself goes on.

CHAPTER 11

The Business Case

In the modern world of business, it is useless to be a creative original thinker unless you can also sell what you create. Management cannot be expected to recognize a good idea unless it is presented to them by a good salesman.
David M. Ogilvy

Of course, if you wanted to be in sales, you would not be reading this book. But to avoid the enormous frustration, not to mention resentment, of developing your network continuity plan only to have it brushed off by management as useless, you must know how to present a compelling case for it. Costs and revenues are the key—and that key, in this context, can include costs avoided and revenues protected.

A business case is, fundamentally, a cost-benefit analysis. If you cannot show that the benefit of doing something will exceed the cost, the business would be foolish to do it. Remember that when you evaluate the cost, you must also include the opportunity cost (the next best use of the money). When you evaluate the costs and benefits, there are two not-mutually-exclusive paths to benefit exceeding cost: increase net revenue and/or decrease net cost. If this seems overly simplistic, I remember the remark of one of my economics professors in graduate school: All the important ideas are taught in the principles classes; the more advanced classes are about being able to prove them.

You must be able to prove your ideas to a naturally skeptical audience. They will have the innate skepticism that we all have when confronted by an idea not our own. But, as senior decision makers responsible for the financial health of an entire business, they must be skeptical of all ideas that allege to change the financial health of the business. If you can show that your idea will probably save money (they already understand that there are no guarantees) and may improve some intangible values as well, you are well on your way to getting what you need.

Understanding Costs

There are three ways to look at costs. All are rooted in economics, but one is not reflected in accounting. The three methods are fixed and variable costs, direct and indirect costs, and explicit and implicit costs. The first two pairs can be summed to provide a measure of total costs; the third pair is where accounting standards and economics diverge. The question you must answer when you present your business case is this: How will total costs change as a result of this plan?

Fixed and Variable Costs

This dichotomy is right out of Economics 101—that principles class most people endure because the university requires them to do so if they want to graduate. Fixed and variable costs are rooted in manufacturing, but they apply in any business; what changes in a service or software firm is how big a part of the cost structure each is.

Fixed costs are those that are incurred regardless of whether the business ever actually produces anything—anything at all. These are costs such as rent, basic utility fees (dial tone on a telephone line, for instance, has a fixed cost to be paid every month even if not one call is made), salaries (as opposed to hourly wages), lease payments on equipment, depreciation charges taken, and so on. Fixed costs have a constant value over the time period being measured, which is no longer than the time period over which the fixed costs cannot be avoided. If this sounds like a tautology, it is. In economics, the short run is the period over which you cannot change fixed costs; the long run is that period of time that is long enough for you to get out of those fixed-fee contracts, change service providers, and so on. The short run is often assumed to be one year (which is taken more as a handy reference value than as evidence that cost structures suddenly change at that particular point).

Many people believe that fixed costs are the same thing as overhead because overhead expenses cannot be charged against any particular piece of work. This is not correct; while overhead is not directly allocated to any particular project, it can often be indirectly allocated (more on that coming up). The difference between fixed and variable costs relates only to whether the amount of the cost changes as a result of a change in product output.

Variable costs are those that do not remain fixed; they increase or decrease, depending on the business's activity. Manufacturing inputs are the classic example; if you don't make cookies, you don't need to buy flour, eggs, and sugar. There are more subtle variable costs as well. Most of your utility costs, for instance, are variable costs because if people are not working in a building, the climate control can be relaxed and the lights can be turned off. As business activity in the building increases, the utility costs will increase, even though they are generally considered overhead. (This is a simple example of why overhead and fixed costs are not the same.)

Increasing output does not necessarily change variable cost by a constant factor, though over small increments of output change you can assume linearity. The change in cost due to an increment in production is marginal cost. The business must pay its fixed costs; therefore, any decision about changes in production are taken based on marginal cost, which is the change in variable costs. Normally, marginal cost assumes no underlying change in the way the business operates, but it is just that which you are trying to change. Therefore, when you present your business case, if you are required to address fixed and variable costs, be sensitive to questions that ask if you are increasing fixed costs. Such an increase asks the business to increase its obligations whether or not it sells any product. If you can, demonstrate that variable costs decline. Increased productivity does this by making people more efficient in producing what you sell.

Direct Costs versus Indirect Costs

Another way to break out costs is by how they relate to a project: Are they a direct result of the project, or are they caused by an effect of the project? Direct costs are immediately traceable to the event or activity in question; indirect costs may be likened to ripple effects. Because of this, direct costs are generally relatively straightforward to estimate, while indirect costs require some assumptions to be taken. A simple example should suffice: Your company is expanding its software development and has decided to produce an integrated office productivity suite (it will have a different set of programs than existing suites). A direct cost of this decision is the increase in payroll of software developers. In order for them to work, you will need new workstations, more cubicles (two people per cubicle is not productive I've been there), another HR and another payroll staffer, a new director, three administrative assistants, more telephone capacity, and possibly more Internet access bandwidth.

These costs are a result of the new project. Drawing the line between direct and indirect cost is where things can get a bit ill defined. The director and the three administrative assistants can reasonably be called direct costs, if the new project is the only thing in the company that they will do. The additions needed for HR and payroll are unlikely to be dedicated to this project; it is more likely that they are needed because the new project plus the existing workload together have made current staffing insufficient. That makes them indirect costs. Likewise, you are not acquiring a new expense in the form of real estate, telephone service, and WAN connectivity, but the increase in each results from the people working on the project and is thus an indirect cost. The demarcation between the two is that indirect costs are the result of direct costs; they are caused by an effect of the project's principal activity.

A direct cost of network continuity will be the additional hardware, software licenses, bandwidth, and personnel required. Indirect costs include such items as the additional real estate required (whether you simply expand an existing location or acquire a completely new one), increased power consumed, the increased air conditioning to counter the heat from the equipment, the additional workload placed on other personnel in the company, and (possibly) a decrease in throughput on some links as more filtering is done or traffic is deliberately fed through bottlenecks to establish control points against hacking. You could also have a negative indirect cost (a reduction in total costs to the company) if, in the process of forcing traffic through controlled chokepoints, you filter out the unauthorized activities that have been going on over the network.

Costs are not always more money out the door. Another indirect cost may be a loss of revenue if, as part of your network continuity plan, you recommend discontinuing a service that you believe poses some danger to the network. Perhaps, as a gesture of community support, you have allowed a nonprofit organization to host its Web site on your server in the DMZ for a nominal fee. Because you do not control the security on that site, you cannot be sure it will not be used as an intermediary in a cyber-attack. To preclude infection of your services and any liability for facilitating an attack, you regretfully end the hosting function. That lost revenue is a cost. Any configuration or maintenance you performed due to their presence, a cost for you, will also go away, so there will be a reduction in both revenues and costs (actual) as well as the reduction in risk (also actual, but harder to quantify).

Explicit and Implicit Costs

Explicit and implicit costs are one of those concepts that is baffling, until it is suddenly clear; there does not seem to be any gradual dawning of understanding on this idea. The simplest way to describe the difference is that an explicit cost can be measured reasonably directly by money actually spent. Implicit costs do not incur actual expenditures; rather, they involve nonmonetary costs. Because of this, and because of the difficulty in quantifying the implicit costs, business accounting rarely mentions them.

Opportunity cost is an implicit cost. You are not actually paying for anything, but you are losing the opportunity to have a possible revenue stream (with its associated costs). Because most of us have somewhat limited discretionary income, when we choose to buy one car, we give up the opportunity to buy a different car. Likewise, when you choose to spend your network continuity money on a contract for on-call support, you give up the opportunity to spend the same money on building out internal redundancy.

Another implicit cost is damaging your reputation in the eyes of customers. If you do not reliably deliver when you say you will, customers will demand a better price in order to compensate them for the risk in dealing with you. That price differential is an implicit cost of your tarnished reputation; proving it exists and determining its size is not a trivial exercise. Being hard to measure does not excuse you from considering implicit costs, at least qualitatively, in your business case.

Valid Comparisons

We've all been reminded, starting in childhood, that for comparisons to be valid, they must compare similar things; apples to apples is the standard we must meet. One problem with business decisions is that they often involve money flows that occur over several years. We have also learned, since childhood, that a dollar doesn't buy what it used to. But price changes are not unidirectional; for what I paid for my (now) three-year-old computer I could buy roughly twice the power I have now.

We have two factors to consider when planning the network continuity: changes in the value of the dollar (inflation) and changes in the price of equipment (which follows the effects of Moore's Law). Equipment price trends are secular (they wobble but maintain an overall direction and an overall rate of change); you can probably forecast them with reasonable accuracy from your experience and through consulting the trade information services. That means that you plan when you will acquire equipment and have a price you expect to pay at that time.

DISCOUNTING, PRESENT VALUE, AND INTEREST

This is not a complicated subject, either conceptually or mathematically. It can be tedious, though, because each year must be treated individually and then the results summed. When we take future monetary values and bring them back to the present, it is called discounting. We are factoring out the expected change in the value of money—all money of this currency—between now and that future time.

For simplicity, let's assume an inflation rate of 10 percent. One year from now, to buy what costs $1.00 today will take $1.10, an increase of 10 percent. Another way to look at it is that, if I have $1.00 today, one year from today, that dollar will buy goods worth only $0.90; my dollar has lost 10 percent of its value.

Assume inflation continues at 10 percent for another year. To buy two years from now, at 10 percent inflation each year, what costs $1.00 today, I will need to spend $1.21 (the price rose 10 percent during year 1, then another 10% from the higher base during year 2). Conversely, my original $1.00 (if I still have it) will now buy goods worth *another* 10 percent less, or worth $0.81 today.

Inflation and discounting calculations use time as an exponent because they use the compound interest formula:

$$FV = PV(1 + i)^t$$

The future value of a quantity of money is equal to its present value times (1+ the interest rate) to the power of the number of time periods for that rate. We usually use an annual rate for *i* and the number of years for *t*. If an expenditure occurs over several years, each year must be brought back to the current one in value.

For example, if I am going to incur costs of $10,000 each year for the next five years, I need to discount each of the four future years back to the present and add them to this year's expenditure to get today's value for that cost. If I assume my interest cost is 6.00 percent, the present value of my nominal $50,000 expenditure is $44,651.06 ($10,000 this year + $9,433.96 next year + $8,899.97 the year after + $8,396.19 the year after that +$7,920.94 in the final year). Answers you get will differ slightly, depending on when your functional calculator assumes the amount is paid—at the beginning or at the end of a given year. I used Excel for this calculation.

In fact, it actually took me more time to break out the individual numbers than it took to calculate the total, using the financial functions built into Excel (and every other spreadsheet I am aware of). Unless you have a burning desire to explore the wonderful world of financial analysis (in which case you need to be reading a different book), use the spreadsheet.

To make a valid decision today, though, we need to take all the costs and all the revenues associated with that decision, whenever they will occur, and report them in dollars (or whatever currency you use) of the same value—measured at a single point in time. That point of time is generally now because that is when the decision is being taken. Therefore, discount all future costs and revenues to their present value.

Once you have all costs and revenues in present value terms, you know the balance between the value of what you will spend and the value of what you will get for that spending. One of the killer apps of the PC revolution was the spreadsheet with financial functions built in; complex transactions spread over multiple years could be analyzed at will, and changes in assumptions (like the effect of an interest rate change in the third year of a multiyear event) modeled with relative ease. You, too, have a spreadsheet, and you can put your proposal in terms the decision makers will readily understand.

Understanding Revenues

Suppose that your network continuity plan will actually require (allow) you to restructure your network to make it work more efficiently. Once the restructuring is done, you will be able to deliver the same work for less cost. That savings could be expressed as a form of revenue.

There aren't really multiple ways to categorize revenues; they are either direct or indirect. Paralleling direct and indirect costs, direct revenues are those changes (positive changes, we hope) that are immediately attributable to the action being contemplated; they are only one step removed. Indirect revenues, of course, are the result of an effect of the action instead of the action itself; they are more than one step removed.

For example, your firm is an NSP, providing a variety of WAN services for a fee. You have SLAs, with the usual penalties of credit for the clients if you fail to meet SLA conditions (a generalization of what is usually a very explicit contract, weighted in the provider's favor by the measurement criteria and the fact that the customer generally has to prove an SLA violation—the provider is not obligated to confess voluntarily). Your network continuity plan may result in winning certain contracts you would otherwise have lost to competitors. Those are direct revenues. It may also, because of your greater reliability, lead to fewer penalties that reduce revenues in future periods. These are indirect revenues; they result from the effect of the plan, reduced failure rates, rather than the existence of the plan itself.

When you assess the increased revenues resulting from your action, remember that, like costs, they must be expressed in today's value. That constant-value expression can be used in two ways: first, to assess if the revenues achieved will, in fact, exceed the costs. Second, you can learn what the break-even value of revenues will be, even if it occurs in a later time period than the cost.

For instance, in the case of the network improvement paid for at $10,000 per year for five years: suppose we believe the change in revenues (both direct revenues and indirect revenues) will be $58,350 in the sixth year. Is it worth making the network improvement? We already know the present value of the costs is $44,651.06 (see the sidebar, *Discounting, Present Value, and Interest*). Discounting the revenues (over six years, not five, because the revenue occurs after the improvement is complete) at the same interest rate (6.00%), we find its present value is $43,602.51—not quite enough. What is enough? The value of today's $44, 651.06 cost in six years, at 6.00 percent will be $59,753.19 (remember, this year is year 0; the sixth year will have an exponent of 5). That is the break-even revenue needed in the sixth year (in sixth-year dollars) for the improvement to pay for itself.

The danger in taking the calculation in this direction is the temptation to find that extra $1,403.19 that would make this a break-even project, even if it requires a little creativity. Had you thought those revenues were there, you should have included them in the first place. It is possible that, with minor changes in how the $10,000 per year is spent, the future revenues might be higher.

But don't be creative with your financials for their own sake. Creativity in finance is not always a Good Thing; remember Enron and its fellows.

Expected Values

When you measure the costs or the revenues that result from an action, you are assuming the action will take place (or has taken place). Yet you also know that many of the actions you need to evaluate will have the associated benefits only if a certain event or set of events occurs. Likewise, when we are reducing costs, the reduction matters only for those occasions when the costs actually do occur. How can we allow for that in presenting our business case?

Some people turn to operations research and apply game theory strategies such as maximin—maximize the minimum value of outcomes among choices (that way, if the worst comes to pass, you at least get the best result of those bad outcomes). There are a number of strategies available, but situations like the Prisoner's Dilemma game show that what is best in long-run, repeated iterations of the game is not necessarily what is best if you are going to play the game only one time.

If you don't want to get caught up in game theory and the complex simulations that often develop (despite good intentions) in operations research, there is a simpler approach, though it may not be as rigorous as the other choices. In most cases, it is good enough. In statistics, it is known as the expected value of a variable. Simply put, the expected value (sometimes also called the expectation) of a variable is the value if it occurs times the likelihood it will occur:

$$E[x] = P_x V_x$$

where P is the probability of occurrence and V is the value assigned to that occurrence.

This is an estimate; not only must you assign a probability of occurrence, but you must assign a value without the event having actually occurred. You may have good reason to postulate that your expenditures will be a certain amount in the event of a disaster that costs you the use of a site. You can project those costs over the time period before you manage to get well (appropriately discounted, of course). But you may or may not get the number right, and the only way to know is to see if the event happens.

Though they are estimates, you can probably develop cost and revenue numbers with some credibility. You can reference the company's prior financials, vendor price quotes you have gotten, and so forth. Developing a probability is less easy. There are some resources available, such as Emergency Preparedness agencies for the probability of natural disasters. You can go to CERT's Web site (the URL is in Appendix A, "References," under *Miscellaneous*) and get the latest statistics on the computer problems to which they have responded. For instance, in 2000, CERT responded to 21,756 incidents, but in 2001 the number was 52,658. That growth is part of a secular trend; we can reliably expect the number to be higher in 2002. From their statistics, you can expect to be attacked; the type of attack and how serious it will be depends in part on what business you are in and how well you have configured your network.

I know of no resource to help you with the probability of a physical terrorist attack, like the one on September 11, 2001. It is possible that the statistics of catastrophes will apply; those are more complex, and statisticians are still developing both the theory and the best methods to calculate them.

> **THE STATISTICS OF CATASTROPHES**
>
> The statistics of catastrophes starts out recognizing that events with vanishingly small probabilities of occurrence that have enormously expensive outcomes need something other than normal distribution-based statistics to assess expectations (of costs, for example). Examples are a 6 mile/10 kilometer meteor striking the earth, a more human 100-year flood, a tornado hitting this particular building, or an earthquake in this particular location at this particular time.
>
> Catastrophes are predicted based on lots of parameters that, collectively, act sort of like a strange attractor: no one of them means much individually, but once they collectively cross a threshold—or they cross a collective threshold (not the same thing)—the event occurs catastrophically: There is no warning buildup. The earthquake is here, the tornado is formed. As an example, one catastrophe theory model predicts grasshopper infestations based on climatic data (mean daily temperature and three-month precipitation accumulations) and historic site-specific information relevant to grasshopper population preferences. The model uses catastrophe theory to predict grasshopper infestations based on the grasshopper infestation level in the previous year and the weather data for the current year.
>
> The insurance industry is paying a great deal of attention to this area (not surprisingly). It estimates that there will be a $75 billion Florida hurricane (dollars of damage), a $25 billion Northeast hurricane, a $72 billion California earthquake, and a $100 billion New Madrid earthquake—but it cannot say when for any of them.
>
> You must choose whether or not to assume such an event will happen in your time horizon. Either assumption is defensible.

Presenting Your Case

You have the ideas now and most of the tools. How do you present your business case?

First, provide numbers and justifications for those numbers. You can first present the cost of various types of disaster, weighted by the probability of each. Then present alternative means of protecting the network from these costs. At the minimum, present an on-call facility approach (such as SunGard) and a network design approach. You may have more than one variation of each of these; present them all or, at least, all those with reasonable valuations (but be prepared to address tactfully some that turned out to be unreasonable, especially if you are doing an oral presentation; your questioners will not yet know the estimated results and may have some ideas that seem reasonable on the surface). Include the costs of choosing each plan, both capital expenses (such as equipment acquisition, possibly real estate) and operating expenses.

Next, address the benefits to the company that will follow from each version of the network continuity plan. Obviously, the benefits will include the costs not incurred if the plan is in place. There should be additional benefits, some direct and some indirect. Benefits are usually revenues, though some may occur in the form of reduced costs. Direct benefits from a network redesign might well include better network performance (if you straighten out some odd growths in the network that developed over time, or if you eliminate nonbusiness traffic when you clean up the hosts). An indirect benefit likely to occur from the redesign is reduced problem response workload on personnel, leading to greater productivity as you spend less time fixing and more time improving. The applicable benefits will depend on what you choose to do in the plan.

In the cost of a disaster (cyber or physical), consider at least these issues:

- Lost business.
- Lost/reduced productivity.
  - Example: Four people for three days cleaning up virus damage, at a fully loaded labor rate of $150/hour is $14,400.
  - Apply to all users affected, including those at other locations.
- Litigation/settlement costs.
  - A hacker gets access to sensitive data.
  - Zombies in your network cause another firm's major loss.
  - Shareholder sues for lack of preparedness.
  - The company fails to deliver the contracted product.
  - Extranet partners sue for failure to perform your part of management/maintenance.
  - SLA costs (if you are a service provider) accrue.
- Physical equipment loss.
- Replacement connectivity.
- Intangible costs, such as damage to corporate reputation.
  - Estimate via cost to repair.
  - Estimate reduction in business.

There is one option that many people forget to include—doing nothing, leaving things at the *status quo*. In all likelihood, it is not a desirable option, but treat it in the same fashion as all the rest. It may be that senior management and/or the board of directors may decide the risk is worth bearing; that is their choice to make. With the cost-benefit structure of this option included in the plan's presentation, it cannot be a choice taken out of ignorance of the likely impact.

You should have detailed numbers, probably in a spreadsheet, supporting the net cost-benefit results on each option. Don't expect people to wade through details to find answers, though. Make it easy for them, and look professional in the process (thereby generating a little halo effect through which your ideas will be taken more seriously). Have a summary section early in your business case that describes the problem you are trying to solve—and remember, network continuity is there to support business continuity. Ensure that your business case evaluation is couched in terms of the *when* and *how* that were developed from the business's continuity requirements.

Large firms will have a standard format in which every business case must be presented. Your readers will know the format and will look at certain sections for the information on which they base their judgments. Don't fight the format; use it to your advantage by placing the most important information early in the appropriate section, where the readers looking for it will find it immediately. Within the format, I recommend the following:

- After the problem statement, briefly—perhaps even literally in 25 words or less—describe each alternative evaluated.
- List the underlying assumptions you have used.
- Have a summary table with the costs of disasters, the probability of each, and the expectation for each.
- Have a summary table of the alternatives considered and their benefits and costs.
- Include the risks you cannot mitigate, or cannot afford to mitigate, and why that is so in each case.

In an earlier chapter, I promised a sort of experiment to demonstrate the value (and operating necessity) of the network to the business. If you can perform it, you will have additional basis for the costs the company must bear in the event of a network disaster. Proposing the experiment depends on the personalities involved; even if they are favorable, you may want to do this informally rather than making it a part of the business case.

Find a senior manager, especially the CEO, and get that person to try this simple thing: "Do your job without any network support for one day, or even half a day. You cannot act as though you know you will have support tomorrow or this afternoon. You cannot slip your project deadlines and arrange to have work to do that can all be done locally." This has to be a day like any other—how much of his or her normal work would the guinea pig get done with no email, no network access (unplug the cable from the wall if you have to), no soft copies of information except what was already (by chance) stored locally? If you want to simulate a virus, the lucky candidate would lose desktop support as well—business the old-fashioned way.

Extend that level of productivity across an entire location or more. What is the cost of doing business that way instead of your normal way? This experiment demonstrates the cost of a network disaster, cyber or physical.

You can choose to insure against that, with your network continuity plan, or you (as a business) can choose to forgo insurance. Having a plan is not mandatory; I think you will find that it is wise.

CDG Example

Let's use CDG's work from Chapter 10, "Returning from the Wilderness," with completely fictitious and conveniently rounded numbers, as an example. Because Shelly Guo is going to look at a five-year period, she assumes the presence of CDG's two expansion locations, Memphis and Fort Worth. Remember that her priorities were as follows:

1. Improve DNS.
2. Improve power redundancy.
3. Improve existing customer alternate routing.
4. Evaluate security (again) based on the changes in the network.
5. Improve the network continuity plan to reflect the possibility of complete loss of one city's NOC.
6. Rigorously exercise the network continuity plan and incorporate the results.
7. Reevaluate the network design for the expansion; will the immediate improvements (#1, 2, 3) affect migration to that design? If yes, decide whether to redesign the future network or reconsider how to execute the immediate improvements.

Within these priorities, we will look at how Shelly addressed the problem of natural disasters. A similar set of alternatives and cost-benefit analyses would be developed for a cyber-attack. Shelly decided that there is no way she can predict the chance of a terrorist attack, so she will assume that their natural disaster preparations will protect against one sufficiently (much as it did for Lehman Brothers after the WTC attack).

Alternatives Considered

Shelly considered these alternatives:

1. No change.
2. Lease a second facility in each city, at least 10 miles/16 kilometers away from the existing facility, and divide operations between them. Each would be fully capable of handling the entire load for that city. Customers would be multihomed to both.
3. Upgrade capabilities at every city to the full suite of CDG services (DNS, DHCP, and so on) and acquire backup connectivity for every customer to an alternate site.
4. Contract with EverReady Services, Inc. (Note: a fictitious continuity provider, like SunGard) for the use of its backup facilities in CDG's cities. The contract applies to every location, and fees depend on the numbers and types of equipment and communications lines required. IP addressing would be preserved, so customers need make no backup plans unless they wish to be multihomed to another provider. They would have to expect an operational outage of the time it would take to reestablish operations, a time to be developed in each city via exercising the plan.

Disaster Summary

When creating the following disaster summary tables, Shelly had to value the damage likely at each site. Being conservative, she chose to assume the complete loss of equipment and facilities in the event of a tornado or flood, and lost business due to power outages from a major winter storm. The former are considered major natural disasters, and the company's insurance covers lost business in those cases; a major winter storm does not receive the same treatment under its insurance. (Note: For simplicity and ease of following the numbers, all costs and benefits are reported in current dollars, rather than being discounted as they should be. This is unlikely to affect the final qualitative outcomes in this instance because we are not positing any difference in the probability of a particular event in any given year.)

Shelly used the basic values of the damages shown in Table 11.1.

Table 11.1 Estimated Total Damage by Disaster Type

CITY	TORNADO	WINTER STORM	FLOOD
KC	$1,500,000	$350,000	$1,500,000
DNV	$1,200,000	$285,000	$1,200,000
MSP	$1,200,000	$285,000	$1,200,000
CHI	$1,200,000	$285,000	$1,200,000
MEM	$1,200,000	$285,000	$1,200,000
FTW	$1,200,000	$285,000	$1,200,000

She then applied the probability of one occurrence among CDG's six cities in a five-year period to find the expected cost of the disaster (see Table 11.2). She found the probabilities of each type of disaster easily enough from the histories available on government Web pages (several URLs are listed in Appendix A). With some digging, she also found the tornado tracks over history and adjusted the probabilities according to their actual office locations compared to those tracks. Similarly, for flooding, she applied the extent of historical floods according to CDG's actual location in each city. She increased the probability in each case by her own "fudge factor" based on the proximity of large water mains (that information was harder to find) and the extent of infrastructure development in the area; the fudge factor reflects the possibility of a man-made flooding event such as what happened in Chicago and Dallas.

She has applied the likelihood of the event occurring at each location; now she has the expected cost CDG will incur if these events happen. The total costs the company faces over a five-year period staggers her. While she recognizes that probability is not certainty, and she knows that she had to be somewhat creative in developing the probabilities of tornado and flooding actually affecting their locations, she did each on its own merit and then applied the result against the damages likely. Once she saw those values, she asked the financially oriented partner for information on the company's insurance policies at each location. Disturbingly, she learned that the insurance was based on the depreciated value of equipment instead of the replacement value. When she warned him that was inadequate, he demanded facts as to why; a depreciated value policy has much lower premiums. She added Table 11.3 to her business case.

Table 11.2 Disaster Summary

CITY	TORNADO	P(TORNADO)	E[TORNADO]	WINTER STORM	P(WS)	E[WS]	FLOOD	P(FLOOD)	E[FLOOD]
KC	yes	.30	$450,000	yes	.75	$262,500	yes	.75	$1,125,000
DNV	yes	.12	$144,000	yes	.80	$228,000	yes	.20	$240,000
MSP	yes	.10	$120,000	yes	.80	$228,000	yes	.25	$300,000
CHI	yes	.10	$120,000	yes	.50	$142,500	yes	.30	$360,000
MEM	yes	.25	$300,000	yes	.45	$128,250	yes	.80	$960,000
FTW	yes	.20	$240,000	yes	.60	$171,000	yes	.20	$240,000

Probabilities estimated (fictitious) for one occurrence that actually affects CDG location in a five-year period.

Table 11.3 Total Expected Disaster Cost by City

DISASTER TYPE	KC	DNV	MSP	CHI	MEM	FTW	TOTAL BY DISASTER
Tornado	$450,000	$144,000	$120,000	$120,000	$300,000	$240,000	$1,374,000
Winter storm	$262,500	$228,000	$228,000	$142,500	$128,250	$171,000	$1,160,250
Flood	$1,125,000	$240,000	$300,000	$360,000	$960,000	$240,000	$3,225,000
Total	$1,837,500	$612,000	$648,000	$622,500	$1,388,250	$651,000	$5,759,250

Based on the possible damage, she thinks CDG may also need to reconsider its operating locations; perhaps there's a reason why property in the flood plain is so much cheaper.

Alternatives Summary

To evaluate her alternatives, Shelly needed the costs of each in order to weigh them against the benefits. As she started to do this, she realized that not every alternative eliminated all disaster costs, so she had to develop the benefits of each alternative as well. Because the plan will be for CDG's entire network, she evaluated the costs and benefits network-wide, rather than on a per-site basis. She also used the depreciated value insurance repayment (CDG's equipment is, on average, 50 percent depreciated now; replacement costs have not greatly changed since the purchase dates) to reduce the costs for one comparison, and the replacement value repayment for another, correcting for the higher premiums of the latter (which her partner had pointedly supplied). She called them *insurance as is* (IAI) and *improved insurance* (II); the latter will cost CDG, network-wide, another $22,000 per year. She also recognized that CDG's insurance in either case did not protect the company from the lost business caused by a winter storm, which made that a cost to be borne unless she could assume that redundancy somehow eliminated it.

Her costs for each alternative, then, had to be calculated twice. Under the insurance as is choice for each alternative, she had to include 50 percent of the lost equipment cost (replacement less the reimbursement at the 50 percent depreciated value), the cost of actually implementing that plan, plus the cost of business lost during a winter storm. When the alternative eliminated lost business, its value in the benefits line balanced the cost and netted to zero. This allowed her to demonstrate continuity's explicit value to her partners. Under the improved insurance choice for each alternative, the higher insurance premiums essentially replaced the lost equipment cost; she still had to include the plan costs and the (possible) lost revenue due to a winter storm.

The following are the costs to implement each alternative named previously:

1. $0 (doing nothing).
2. $3.6 million in capital expenditures (as they redistribute some equipment between the old and new locations in each city) and an extra $750,000 in total operating expenditures over the five-year period (redundant locations in each city).

3. $0 capital expenditures and an extra $3.2 million in operating expenditures over the five-year period (redundant connections between each customer and two CDG locations).

4. $0 capital expenditures and $50,000 in total fees over the five-year period, plus actual usage; she estimates CDG will likely need to use the on-call facility once per city in the five-year period, at a total cost of $320,000 for the usage period (on-call facility).

The benefits under each alternative were the reduction in lost business costs due to winter storms (if applicable), and intangibles, such as validating the company's reputation for reliability. She has no idea right now how to value them; she plans to get some guidance on that from her financial partner. In the meantime, the explicit costs and benefits are shown in Table 11.4.

Shelly concludes this section of her business case with three points:

1. Regardless of the alternative taken, CDG needs to revise its insurance. The full-replacement-value insurance is preferable in every case.

2. The best alternative is #4, to contract with EverReady Services, Inc., for a backup facility. If CDG is committed to staying in the market, and so must have continuity, the next best alternative is #3, a fully distributed network functionality (much as it pains her to give up the close control she has exercised to this point) with backup connectivity for customers.

3. Alternative #1, with improved insurance, would actually be better financially, but because she could not yet cost the damage to their reputation in the event of a major outage, she does not believe that result will stand.

Risks Not Mitigated

This section of the business case is often forgotten, but it is important to be clear about the risks you cannot, or choose not to, mitigate and why. In the case of CDG, Shelly identified two situations. First, she specified a terrorist attack, such as a bombing or even the use of a CBR weapon or an EMP attack. The probabilities of these are impossible to calculate, and she decided that in the event of such an attack, customers will have more pressing concerns than connectivity.

Second, while Memphis is an attractive location for expansion, it is just over 100 miles/165 kilometers from the town of New Madrid, Missouri.

Table 11.4 Alternative Comparisons

ALTERNATIVE	1 IAI	1 II	2 IAI	2 II	3 IAI	3 II	4 IAI	4 II
Benefits	$ 0	$ 0	$1,160,250	$1,160,250	$1,160,250	$1,160,250	$1,160,250	$1,160,250
Equipment replacement	$2,879,625	$ 0	$2,879,625	$ 0	$2,879,625	$ 0	$2,879,625	$ 0
Insurance	$ 0	$ 22,000	$ 0	$ 22,000	$ 0	$ 22,000	$ 0	$ 22,000
Lost business	$1,160,250	$1,160,250	$1,160,250	$1,160,250	$1,160,250	$1,160,250	$1,160,250	$1,160,250
Plan implementation	$ 0	$ 0	$4,350,000	$4,350,000	$3,200,000	$3,200,000	$ 370,000	$ 370,000
Total	**$4,039,875**	**$1,182,250**	**$7,229,625**	**$4,372,000**	**$6,079,625**	**$3,222,000**	**$3,249,625**	**$ 392,000**

IAI = insurance as is; II = improved insurance.

When the earthquake eventually occurs somewhere along that fault, Memphis is likely to sustain extensive damage. Because there is no predicting when such an event will occur, and because customers will again have their own survival to look to, she has decided to not attempt to protect against it, either.

Finally

Even the best business case does not guarantee that your network continuity proposal will be adopted. Although it is the most economical proposal, if the money to pay for it is not there, the company may simply have to bear the risk of not implementing a continuity plan. Management may also decide there are better alternative uses for the money (and your instructions in either case will be the equivalent of "just do the best you can"). Likewise, the best plan for you will not look like the best plan for another company, even one in the same line of business. You may use a different discount rate, you may have a different network evolution plan, or your board of directors may have a very different risk tolerance than another company's.

Finally, even in the business case, you will have to recommend compromises. The network that is easiest to secure, for instance, is not necessarily the best design to be redundant or flexible under physical stress. (What happens if one of your choke points is where a failure occurs?) You may need to trade off designing circuit-like connections, where you realistically need to secure only the endpoints, with the more flexible and throughput-efficient system of securing each packet wherever it goes in the network.

Your choices will be based on your circumstances and your preferences. You will most likely optimize rather than maximize; do so according to criteria you consciously choose and explicitly state in advance in your assumptions. Then model the outcomes against what-ifs. You may not be able to model quantitatively, but you should be able to do at least a qualitative analysis.

It will be a larger task than you think at the beginning, as more and more factors to consider become apparent. But the effort will only improve your network's reason to exist—its support of the business.

CHAPTER 12

Conclusion

Honor isn't about making the right choices.
It's about living with the consequences.
Midori Koto

The pundits and talking heads told us all how the world changed on September 11, 2001. As one who was bombed by terrorists more than 20 years ago, I would disagree. The world has not changed—but certain aspects of it are both more violent and more willing to bring that violence to the United States than they once were.

In fact, businesses with overseas operations have been dealing with violence for years, especially in the developing world, but also in those countries that are well on the way to full development and straining in the process. The degree of violence reflects in part the social tensions that accompany a fear of being left out of the chance for a better life. With ubiquitous communications to emphasize differences and facile travel to eliminate barriers, the threat of socio-political violence is likely to be with us for the foreseeable future, wherever we are.

Why have businesses stayed with their operations in places where they are subject to violence, perhaps especially targeted for the violence if they are easily identifiable as American? Simply because, despite the greater costs, it remains worth it. Just as in your business case, the benefits outweigh the costs. The benefits need not be entirely financial or directly financial. India has struggled with terrorism from diverse sources over the years, but it has a growing population, many of whom speak excellent English and have solid educations. Software developers who learn in India's schools learn on old computers with limited memory—and they write well-crafted, tight code as a result.

The benefits of continuing overseas operations, then, can include a market presence in a country with excellent resources—human as well as natural—and the resulting greater productivity. The costs are relatively low, especially if the product can be created in the low-cost area and transported readily and cheaply to the high-priced market. A well-respected investment banker in Hong Kong has warned that businesses that do not operate much informational product development in these low-cost/high-productivity areas will not be able to compete. Indeed, information services is the epitome of such a business. But whether your company's product is information itself or a physical output like steel, *your* job is delivering the network that makes it happen, wherever it needs to happen. That makes your operation a part of the rapidly growing service sector, which now constitutes between 60 and 70 percent of the global gross domestic product—the value of all new goods and services created worldwide in any given year.

If the information transfer of product and cooperative work is free, so are other transfers. Those disaffected by religious/ethnic/social/political philosophy/other cause differences may be among those your local manager hired, or they may simply hack into a less-secured part of your network. Now there is a free ride into the rest of your network, especially if they can compromise a host and use it to enter at will. The spread of technical education does not stop at a moral boundary; like any other part of science and technology, networking expertise can be used for good or ill.

Necessity

Of the two principal types of disaster—physical and cyber—the latter is far more likely to occur. In fact, if you will recall the CSI/FBI Computer Crime Survey, you are likely to have already been attacked, whether you recognized it at the time or not. You must be prepared to handle an attack, at any moment, in a fashion such that your network's ability to perform its business function—the delivery of information from a place of lower utility to one of higher utility, thereby increasing its value—is not at risk. The attack may crudely disrupt all network operations, thereby announcing to everyone connected that this network is compromised. More dangerously, the attack may be subtle and completely unnoticed, pilfering your company's most valuable property—its proprietary knowledge—or stealing customer information, such as account numbers and passwords, credit card numbers, and more.

You must protect yourself against a range of attackers, from voyeurs and thrill-seekers all the way through those out to see how much they can change your network before you notice them, and on to the subtle and skilled, who do their best to leave you no idea they were ever there. Protecting against the final group requires proactive defenses, such as keeping unalterable logs and reading them, and knowing the characteristics of your network's traffic so well that an alteration in the pattern attracts your attention.

Cyber-attacks have a secular trend of increasing every year. The attacks themselves become more sophisticated because the methods are shared. Once developed by one, they are usable by all, and some among the all make marginal improvements, which are also shared. Like every other conflict (or arms race), there is an interactive evolution by the offense and the defense. Your peaceful, strictly business network becomes more and more important to the business's ability to even be in business, and so it becomes a more attractive target for the hacker. In response, you make it a harder target. That makes you a greater challenge, and the cycle continues. Ideally, you would prefer to be invisible or, if noticed, dismissed as being too boring to bother with; you would break the cycle, at least with respect to your company's participation in it. But there are thousands, probably even millions, of hackers (they don't register with an agency, of course), and you will never manage to bore them all.

You have no choice but to defend against them.

Basic Defenses You Must Implement

Most cyber-attacks exploit known holes and weaknesses in operating systems and applications. The very first thing you must do is ensure that every OS and every application on every host has the most current software (including patches) for every business purpose it serves *and no more*. You have a multitude of hosts on your network—desktops, laptops, servers, storage devices, printers (remember, if it talks to the network, it's a host and it's exploitable), Web servers. Most of them probably have the default installation of their operating system and every application—that's the easiest way to go, and most people don't know better.

Look again at the Top Twenty Vulnerabilities. Most of them arise from default installations and failure to patch well-known flaws in software. Whether those flaws should be there is irrelevant to your problem. They exist, and you must deal with that fact. Once you have corrected these basic problems, which may be found in every network, you can begin to use your architecture to help protect you further.

Why not secure the perimeter first and then correct the internal flaws at a manageable pace? There are two reasons. First, having fortified the perimeter, some may feel secure enough, thank you, and see no need to put any effort into repairing the hosts—and it will take effort, lots of effort, that will leave users irritated and angry when their favorite little features are taken away. If some of those users are senior management, you have a serious internal sales job to do. The second reason is interrelated with the first: All those hosts are still wide open, and many attacks originate from the inside. It is sad, but it is also true: You can't always trust your own people.

And if outsiders do get in past the perimeter defense, they are home free.

The Deck Is Stacked Against You

If only it were not so easy to mount a cyber-attack. The protocols that deliver the information wherever it needs to go, making their presence truly ubiquitous, are porous, to be charitable. To also be fair, they were never designed with this environment in mind (literally—no one imagined the kind of networking and internetworking that is taken for granted today when these protocols were designed). We will all be playing catch-up for a long time.

Can you run farther and faster facing forward or backward?

Silly question, isn't it? But you must always be at least turned, looking backward, while you try to move the network forward. You cannot just abandon legacy hardware and software that is still good enough to do the job *and is already paid for*. Backward compatibility will be an albatross around your neck because the existing infrastructure simply will not be abandoned, meaning whatever you evolve to must be compatible with your company's installed base. That will make it less capable than it would otherwise be, but it is nonetheless the better business decision for reasons of cost-effectiveness.

Because you must deal with porous protocols in a less-than-perfect network architecture (hardware and software, as well as logical topology), you must take a proactive approach to security. Your network's performance, and the integrity of the information that flows over it, will depend on a defense in depth, and that depth includes your hosts. It will also include monitoring and management; you will need to practice being able to track an intruder through the network. Remember that fox-and-hounds may be a game to the hounds, but it is survival to the fox—therefore, who tries harder?

You and your people will have to be smarter.

And we have not yet really seen a cyber-attack used as a weapon of terror, but it is only a matter of time before we do. What process controls in your company's production are automated? What could go wrong if a hacker got in and changed the command set? If nothing else, how would you pay the criminal fines (not to mention the class-action civil claims)? Depending on the number of casualties, it is entirely possible that penalties would not be limited to financial punishment for the incompetent managers who allowed such an event to occur.

Catastrophes Happen

Here's another "if only": If only we could know when catastrophes will happen so we could be prepared. Unfortunately, predicting the occurrence of catastrophes depends on accurately identifying the associated factors (not necessarily causal factors; the best that may be said in that vein is that such factors appear to be contributory) and then measuring them with sufficient precision at the correct times to yield a prediction before the event.

Predicting after the fact is a lot easier.

While the insurance industry is leading the way in developing the factor associations and measurement criteria, they are not yet to the point of venturing more than the broadest of predictions. Yet predicting disaster is their business, for that is the primary input to their premium calculation. Smoker or nonsmoker? Heart disease in the family history or not? The factors on which to base a life insurance premium are evolving as our health profile evolves. The factors associated with catastrophes are much harder to develop because the number of events to analyze is so very few, statistically speaking.

With natural disasters, we have the beginnings of danger profiles. You would be well advised to know where you are at risk from natural disasters and to ensure through your architecture or a contracted backup service that, when it happens, it will not damage your business more than is absolutely unavoidable. The same preparation, with a security twist, will help you defend your network's function against a terrorist's physical attack or even against plain old bad luck.

Unnatural disasters, though, evolve. For those whose technical bent is less mathematical and more physical, the offense-defense struggle is the same as it has ever been. Bombs become more sophisticated and bomb detection does, too. Biological terror is threatened, and public health agencies improve monitoring and medical practitioner awareness. Further, target selection evolves to make the best political statement. Al Qaeda did not just attack entities of the American government; it also struck at the symbolic heart of our economic system. If your company has an operation overseas, especially if you are recognizably American, you can be a focus of a physical attack in order to strain the local government's relations with the United States, as well as for the sheer delight of shooting at the eagle from a distance. Your people and facilities were just a means of making that political statement.

Knowing that does not mitigate in any way the damage done to your facilities, equipment, and people. And it does not magically restore that portion of your network to its role of business support.

While catastrophes are discrete and discontiguous events, network survivability and continuity is a process flowing around those events. The process will have periods in which it is less effective, after which you (or, unfortunately, your successor) will gain access to more and/or better resources. Once they have been applied, your company will have periods of more effectiveness. Again unfortunately, it is in those periods that senior management sees less need to keep improving network survivability and continuity, which exposes the business to the newly improved attackers yet again.

You cannot do this by yourself, either. You will have to devolve responsibility for some things to others. If you give them the responsibility, they must have the commensurate authority, else how can success or failure be theirs? Of course, you remain responsible for their performance, so devolution does not totally relieve you of responsibility for outcomes, only for the direct creation of them.

Your Recovery

It will happen. Perhaps it has already happened to your company once, and you are determined to do better the next time. It may be hard emotionally to accept the idea that another catastrophe could strike or another worm as devastating as Nimda could tear through your network again, but logic can find no reason why you should suddenly become immune. Perhaps a darker sense even prompts you that, once it happens, the virtual jackals will be circling your company's wounded network, looking for another chance to dash in and nip off a piece.

You survived this far; now what?

"Now what?" must be a part of your planning as well. A plan that gets you to the other side of the river, dripping, triumphant, and empty handed with work to do is not much help—especially if the plan never mentioned that you'd need all those tools you let go of when you plunged into the current. The questions you ask when you plan network continuity must include what you will need to do and how you want to be able to do it, once survival is ensured and it is time to rebuild, recover, and even move forward. You must ask what you will be able to do given what you will have to work with at that point.

You are planning for more than mere survival; you are planning for success despite disaster. No plan survives contact with the enemy, Clausewitz said. Your plan will never be an exact fit for what actually happens. But if you manage a loose fit you have something to work with. Plan how you will work with differences from your expectations; think about possible adaptations. How will those adaptations affect the shape your network and your business are in when the crisis is over and it is time to go forward?

Don't plan to go back. Plan to go forward, and, cold-blooded as it may seem, take advantage of the disaster to become better than you would otherwise have been (a sort of "living well is the best revenge" attitude). Backward compatibility was alluded to earlier as an albatross around your neck. After the disaster may be your chance to cast it off and make serious improvements to your network, improvements that you could not have made before and remained compatible.

And here is a final, planning-orientation morsel for thought: If your business chooses internal redundancy via your architecture, step back and look at the virtuous cycle of development you can make that become. As, for instance, you bring Latin American network operations up to capacity to back up North America, Latin American business operations will take advantage of the greater capacity and improve business. Now, to back up Latin America, you may need to revisit your European network and improve it, lest it not be able to help North America carry the load if something happens to your major node in Santiago, Chile. Likewise, Eastern Asia is growing while the rest of the global economy pauses for breath; you are racing to keep up with your network needs in Asia, but you can use the opportunity to scale that topology for improved global performance, not just regional. As each region improves, you evolve the global network to become more robust, a necessity because the threat continues to evolve as well.

But the only way you can have the time to do this is to let go of more of your daily supervisory roles. You will develop others and do more for the business as a whole, which allows you to make the others' development more beneficial to the company and to them because they learn on better projects. You have created another virtuous cycle.

Spreading the workload not only develops them professionally, it forms a kind of insurance for you and the company, too. You do not have to do everything if you have capable people who can do it instead. That yields precious time for you to actually pause a moment and think about what might come your way next; even a little forethought is generally better than none. As for protecting the company, suppose that when disaster strikes, you have just departed Frankfurt on a nonstop flight to Chicago (or, worse, Tokyo to Chicago)? How could you direct things and make it all happen when you are not there and not in communication?

Then, too, suppose you are one of the casualties. Business continuity is about no one event or no one loss crippling the company's ability to survive and prosper. When an organization depends on networked information to perform its functions, a working head of IS may be more critical to continuity in some cases than the CEO and the entire rest of the senior management combined. If you are in the hospital unconscious, there had better be someone who could act on your behalf both professionally and personally.

All of us have a bit (or more than a bit) of Shelly Guo in us: It is hard to let go of the things we cut our teeth doing and use that freed time to do our real, management job. But letting go gives us the time to think about our choices and their results. That is even more important because some choices, once taken, are unrecoverable. The scrambled egg cannot be prepared sunny-side up, ever. A mistaken architectural trade-off will haunt your future work—or, more likely, your successor's.

Trade-offs

There are no easy answers to some questions when you plan for network continuity. Survivability requires trade-offs; those trade-offs will need to be taken consciously, while remembering that you are, in effect, creating a system. It will have those systemic properties we discussed in Chapter 4, "Murphy's Revenge": complexity, interaction, emergent properties, and bugs.

Systemic Behavior

Your survivability system will be complex because your network is complex, and you must retain that complexity to preserve its functionality. The problem for you, of course, is that you are trying to preserve a certain complex behavior with a significantly different set of elements. That is why, in planning for continuity, it is important to focus on the services provided, rather than how they are currently provided. Remember the Twenty Questions (plus a few more) in Chapter 6, "The Best-Laid Plans"—they are intended to make you think about the services your network provides to the business and how those can be replaced in the event of a natural or unnatural disaster. Likewise, you must be able to quarantine corrupted services from contaminating still-valid ones in the event of a virus or worm attack.

The replacement elements will interact with the surviving elements of the original network in new and interesting ways. Murphy's Law suggests that not all of those ways will be delightful ones. Yet you will know of their existence in advance only if you exercise the plan; a truly operational exercise would smoke most of those surprises out (there will always be some that simply take time to appear), but you are unlikely to be able to actually divert real, revenue-generating traffic to see what effect it has on things, just in case. The direct implication of these facts is that, in your exercises and simulations, you must be careful to walk through the steps involved in something, rather than just assuming such-and-such an activity comes up as expected (or because the plan intended it to).

Emergent properties of systems are also difficult to plan for. If we knew they were coming, they wouldn't emerge unexpectedly, would they? And yet they emerge from the planned activity of the system as opposed to a failure in it. Anticipating them requires that you and your most thoughtful and creative people step back and look at the replacement system (the one that would exist after the plan's activation) with fresh eyes. Recall that Lehman Brothers had an excellent plan ... until it realized that its operation in the hotel needed far more connectivity than a hotel normally needed to provide for its guests, even business guests. Trace the service threads in your replacement system, and include communications as a thread (or, possibly, a mega-thread composed of many subordinate services)—where are the bottlenecks in those threads? Those are where unexpected behaviors are likely to emerge because that is where the interactions will be concentrated.

And, of course, for want of a nail the kingdom was lost. It wasn't that simple in the original quote: Losing the nail caused the horse to fail, which caused the rider to fail, which caused the battle to be lost, and defeat cost King Richard III his head. A little bug in the preparation led, through a cascade of consequences, to the final one entirely out of proportion with the original problem. There will be bugs in your plan, things you just plain didn't get right. Wouldn't you rather find them in an exercise than during the real thing? That in itself is an argument for exercising the plan, at least at a CPX level.

Even then, like emergent properties, some bugs will show up only when real operations commence with the altered system. They will be fewer, by the number you corrected, and your team will be practiced at finding and correcting them. The familiarity they will have gained with the network structures in the course of the exercises will help as well. Consider those exercises preventive maintenance on your most important recovery system: your people who will be on the spot when it happens, whatever it turns out to be.

As you examine the expected systemic behaviors of the altered system resulting from your plan, you can try a kind of rough sensitivity analysis. What happens to the behaviors and performance if we do less of this and more of that? You must ask not only how do things change, but what difference in the necessary services do those changes make?

Standardization versus Resiliency

Part of network survivability is designing things not to fail in the first place. While there is not much you can do about a natural catastrophe (if your corporate headquarters is in the San Francisco Bay Area, sooner or later there will be a major earthquake, and headquarters will be out of contact for a while, at least) or a terrorist event (especially because the warning indicators are so much more persuasive after the fact), you can do a great deal to make your response and recovery easier if all workstations essentially look alike—all Windows 2000, all Solaris, all whatever. Likewise, it is far easier to secure the operating system and the applications on every host if support personnel have to know only one OS and one set of applications, from a single vendor.

One slip in such a scheme, though, is a slip that can be found everywhere in your global network.

The alternative is to use a mix of incompatible operating systems (and therefore application programs, probably from different vendors). This may have a higher total licensing cost, depending on the economies of scale in a particular package. It has the advantage, though, of leaving you less vulnerable to a cyber-attack *if* you make the investment in your people to be competent in multiple systems. For instance, if your primary DNS authoritative server runs UNIX and your secondary runs Windows 2000, an attack that can cripple one is unlikely to have any effect on the other. A virus that is propagated through Outlook will not affect UNIX hosts, but the opposite is true of one propagated through sendmail.

If you are not willing to make the requisite investment in your people, whether for purely financial reasons of price or because turnover in qualified personnel is too high, then the resiliency approach actually leaves you *more* vulnerable because there are now more possible exploitable paths into your network. Your firm has to choose which you will do, for reasons that make sense in the context of both your business and your financials. This is one of those choices that, once taken, determines your network's path for a very long time.

Pay Me Now or Pay Me Later

Of course, because we are dealing with probabilities, it is possible that later will never come. Does that mean all the money you might spend on network continuity could be a waste, and it therefore ought not to be spent? After all, if it makes things a bit more inefficient (by creating choke points to facilitate cyber quarantine) or expensive (because redundant facilities have some excess capacity), then we have reduced profit. In every market, bull or bear, companies with better profits have better share prices. If a large part of management's compensation is in stock options or bonuses triggered by a certain stock price improvement, there is a strong incentive not to spend any extra money, anywhere.

Who could really blame the company for a massive hurricane? Who could have blamed Lehman Brothers for failing to be operational the next day (or even the same day, for settlements)?

After Hurricanes Andrew and Hugo, we know better: No business in a hurricane-risk area should fail to be prepared to continue operations without the facility in that area. Likewise, after the WTC and Pentagon attacks, no business anywhere can be excused for not planning for the possibility of disruptions due to terrorism, physical or cyber.

So your business certainly could be blamed, in court as well as in the stock market, by anyone who lost money as a result of your disrupted operations, including the shareholders whose asset value sank dramatically.

Your company will have to decide which risk to bear: the risk of a catastrophic financial loss that may or may not occur or the certainty of a reduction in profit year after year with no tragedy conveniently at hand to make it obviously a matter of the board's strategic foresight. Not all of the Exchange's members thought the NYBOT was wise to spend $300,000 per year for five years on backup support from Comdisco—until, one fine day in September, they suddenly needed it with no warning whatsoever.

This "later may never come" argument is why, whenever possible, you should make the network's survivability a part of your long-term architectural evolution and your continuity plan the least-expensive choice to fill the gaps that might occur between now and then. In a sense, this is a specious argument, for your network's evolution will never end and new gaps will appear as you go along. But this approach takes the long-term view that you will ultimately support yourself and eventually will not waste money and other resources on payments to have a capability you hope you never need. At the same time, it is short-term efficient in that it does protect the network's business function as affordably as possible.

If your business depends on real-time access to constantly changing data, as does the financial services industry or many portions of the health care industry, you can readily justify a migration to a more survivable architecture, even if you must bear more risk than you are comfortable with in the meantime. Without continuity in the face of disaster, you simply cannot deliver the business. Plus, in the process of improving your survivability against cyber-attack by cleaning up poor installations and removing unneeded services from the multitude of hosts you must maintain, it is entirely possible that the existing network will run more efficiently on its current resources. What's more, your personnel, *once the cleanup is over*, will spend less time fixing problems that never should have arisen. That's improved productivity and an improvement in profit by reducing your current operating expenses.

Perhaps you do not need that level of responsiveness; an on-call service is sufficient for your operations. Do not expect to be able to write that contract after the disaster has happened: The line at the service company's door can become very long, very quickly.

Choose the alternatives that work for you, based on a careful and well-thought-through review of what you need and when. But remember as well that not making a choice today leaves you one less day in which to become prepared. Firms that had planned to clean up their networks starting September 19, 2001, were one day too late; Nimda had infected over 100,000 computers by the end of September 18, and it was spreading rapidly. Those who planned to evaluate alternatives for physical disasters after they had their third-quarter financials in hand ran out of time on September 11, 2001.

When you do take your decision, document the reasons. There will come a time when the choice will have to be taken again; if conditions have changed, knowing why the existing choice was taken may help clarify whether, in the new and different circumstances, it remains the best alternative. Plus, living as we do in a litigious world, your documentation will show that you did consider alternatives, and that your company made a business decision for business reasons, not greed or thoughtless disregard for those whose interests the board of directors should have served.

If you fail to take a choice, you have chosen to be unprepared. You will find that hard to defend when you do meet the consequences. It is only a matter of time before something, probably not quite what you expected, causes you to need a plan in place. One had best be there.

References

Books

Allen, Julia H. 2001. *The CERT Guide to System and Network Security Practices.* Boston: Addison-Wesley.

Dupuy, Trevor N. 1977. *A Genius for War: The German Army and General Staff, 1807-1945.* Englewood Cliffs, NJ: Prentice-Hall, Inc.

Ellison, R. J. et al. 1999. "Survivable Network Systems: An Emerging Discipline." Technical Report CMU/SEI-97-TR-013 ESC-TR-013.

McCarthy, Mary Pat, and Stuart Campbell. 2001. *Security Transformation: Digital Defense Strategies to Protect Your Company's Reputation and Market Share.* New York: McGraw-Hill.

Messenger, Charles. 1976. *The Blitzkrieg Story.* New York: Charles Scribner's Sons.

Power, Richard. 2000. *Tangled Web: Tales of Digital Crime from the Shadows of Cyberspace.* Indianapolis: Que Corporation.

Power, Richard. 2001. *2001 CSI/FBI Computer Crime and Security Survey.* San Francisco: Computer Security Institute.

Stoll, Cliff. 1989. *The Cuckoo's Egg: Tracking a Spy Through the Maze of Computer Espionage.* New York: Pocket Books.

Taylor, Telford. 1979. *Munich: The Price of Peace.* Garden City, NJ: Doubleday & Company, Inc.

Wadlow, Thomas A. 2000. *The Process of Network Security: Designing and Managing a Safe Network.* Boston: Addison-Wesley.

Web Sites

The URLs listed in this section are valid as of February 2002.

Disaster Planning

http://www.fema.gov
   U.S. Federal Emergency Management Agency home page.

http://www.disasterplan.com/

http://www.disasterrecoveryworld.com/

http://stone.cidi.org/disaster/
   International disaster situation reports.

http://www.nap.edu/books/0309063604/html/
   Mileti, Dennis. 1999. *Disasters by Design: A Reassessment of Natural Hazards in the United States.* Washington, DC: Joseph Henry Press. Text of the book available free online.

Earthquake Hazard

http://quake.wr.usgs.gov/
   Earthquake hazards in the Northern California/Nevada region.

http://www.ngdc.noaa.gov/

   U.S. National Geophysical Data Center home page, links to World Data Centers.

http://www.abag.ca.gov/

   Association of Bay Area Governments home page (San Francisco Bay Area—links to local earthquake hazard information as well as much other useful data).

http://www.usgs.gov/

   U.S. Geological Survey home page (links for many natural hazard information sets).

Other Government Information (U.S.)

http://ftp.fcc.gov/oet/outage/

   FCC outage reports.

http://www.nipc.gov/publications/highlights/highlights.htm

   "Highlights" publication of the National Infrastructure Protection Center.

Miscellaneous

http://www.internethealthreport.com/

   A site with real-time data on Internet connectivity quality (click on "about this site" at the bottom of the opening page).

http://www.menandmice.com/dnsplace/healthsurvey.html

   Information on DNS server location problems [see Chapter 8, "Unnatural Disasters (Unintentional)"].

http://www.cert.org/

   Home page of the CERT (Computer Emergency Response Team) Coordination Center.

http://www.nwfusion.com/news/1999/0705gigabit.html

   Article on network bottlenecks and throughput issues.

http://www.computeamtx.com/services/disaster_recovery/

Disaster planning firm; site has several photos of client offices damaged by Ft. Worth tornado—a graphic illustration of the damage you might face.

http://www.arin.net/

American Registry for Internet Numbers; allows you to identify ownership of IP addresses, also has excellent library with background information on how ISPs work, links to other registries; serves the Americas, the Caribbean, and Sub-Saharan Africa.

http://www.iana.org/

Home page of the Internet Assigned Numbers Authority.

Natural Hazard Costing

http://sciencepolicy.colorado.edu/sourcebook/

An excellent source for damage estimates by type of disaster and by state.

http://www.colorado.edu/hazards/sites/costs.html

More on costs of natural hazards.

Terrorism

http://www.st-and.ac.uk/academic/intrel/research/cstpv/

The Centre for the Study of Terrorism and Political Violence at the University of St. Andrews.

http://www.ict.org.il/

The International Policy Institute for Counter-Terrorism.

http://www.terrorism.com/

The Terrorism Research Center (global information).

http://www.fas.org/irp/threat/commission.html

Report of the Bremer Commission (National Commission on Terrorism).

http://www.usis.usemb.se/terror/index.html
   Global patterns of terrorism reports (annual).

http://www.fas.org/nuke/intro/nuke/emp.htm
   Information on EMP.

http://www.stimson.org/cwc/ataxia.htm
   A report on the chemical and biological terror threat in the United States.

http://www.state.gov/s/ct/rls/pgtrpt/2000/
   U.S. State Department report on patterns of global terrorism, 2000.

http://www.grubb-ellis.com/research/WhitePapers/NYCAttack.pdf
   Report on the extent of real estate damage done to the World Trade Center and its surroundings.

UPS Capabilities

http://www.jetcafe.org/~npc/doc/ups-faq.html#0204

http://www.pcguide.com/ref/power/ext/ups/

Volcanic Eruption Data

http://vulcan.wr.usgs.gov/Vhp/C1073/disaster_averted.html
   The alerts and their timing during the Mt. Pinatubo eruption.

http://vulcan.wr.usgs.gov/Volcanoes/Philippines/Pinatubo/description_pinatubo.html
   Eruption data.

http://wrgis.wr.usgs.gov/fact-sheet/fs036-00/
   Mt. Saint Helens eruption data.

Weather Planning

http://weather.unisys.com/hurricane/
   Tropical storm information.

http://www.nws.noaa.gov/

U.S. National Weather Service home page.

http://www.noaa.gov/

U.S. National Oceanic and Atmospheric Administration home page (major storm information).

Questions to Ask Yourself

In Chapter 6, "The Best-Laid Plans," we discussed the questions you need to ask about what your network does, for whom, and perhaps how you might be able to replace that service today, with resources already in hand, if a major location were completely out of the picture. These are questions you use to characterize your network and its response (including that of the people who operate it) if a disaster happened today.

The beginning of your planning is to know what you have and how it currently works (or doesn't work, as the case may be).

- What capabilities are critical?
- What facilities provide those capabilities?
- Will the entire capability need to be replaced (is partial function possible in some situations)?
  - Can the missing portion be provided from another location, or must the functions be colocated to work properly?
- Can other, existing facilities provide those functions now?
  - Is this one-for-one, one-for-many, or one-for-all redundancy?
  - Can they support more than one failure (or do you wish to assume that is too unlikely)?
  - Is it a collective redundancy (workload distribution of one failed location among all others)?

- Does everyone know who will get what?
- Does the redundant location have the information needed to assume the function?
    - Can it get the information in a timely fashion in a crisis?
- Does everyone know how to activate the redundancy?
    - Is there a procedure established?
    - Is a particular person's/position's authorization required?
    - Does that person/position have a backup for vacation, incommunicado, and so on?
    - Does the backup person know when he or she is *it*? How?
    - Has it ever been tested?
        - Were the problems fixed?
        - Was it retested to check the fixes?
        - How long ago was that?
        - What has changed since then?
- Is there a critical person?
    - Is there one key person who can always make things work?
    - Is he or she always available?
    - If he or she left the company, would the critical knowledge leave, too?
- What is the probable cost if you do not provide redundancy?
    - What are the direct costs (Service Level Agreement penalties, for example)?
    - What are the indirect costs (lost contract renewals, lost new business, litigation, and so on)?
- What alternative means of providing the capability can you acquire?
    - How much does each cost (initial and lifetime expenses)?
    - Do they enhance ordinary operations in any way to offset some of that cost?
        - Can they be made to do so?
- How long can you operate in this mode?
    - What are the operational effects of going too long?
    - What are the personnel effects?

- What will your customers think?
  - What will you tell them?
- When operating from redundant facilities, what becomes the new critical capability? (Do *not* assume it remains the same.)
- Who must be notified internally?
  - Who will make that notification?
  - Who is the backup for making and receiving notification if the primary is not available?
- Who must be notified externally?
  - Partners.
  - Customers.
  - Vendor support. (Be first in line for recovery by being first to notify...)
  - Do both the primary and the backup notifier have those numbers in portable devices?
- Do emergency services know anything about your facilities?
  - Hazardous materials like battery backup banks and legacy fire suppressants.
  - Presence and location of DC power sources (inverters, and so on).
  - Locations of substantial electrical equipment.
  - Locations where the last personnel out of the building would probably be.
  - Their probable egress routes.

APPENDIX C

Continuity Planning Steps

In Chapter 9, "Preparing for Disaster," we looked at creating a network continuity plan. This can be broken down into logical processes, each of which has a number of steps. There were figures associated with some of these steps, simply to provide a visualization of things like services decomposition.

Network Requirements

Before you can plan your continuity, you must understand what the network must deliver, no matter what.

- Define the network's mission.
- Decompose the mission into mission-critical services.
- Define any mission-critical attributes for each service.
- Perform bidirectional service traces to locate any critical elements/single points of failure.

Threat Analysis

Once you know what must continue, you need to understand what might cause a disruption.

- Evaluate network threats.
  - Physical.
  - Cyber.
- Apply those threats to your network, with special attention to your critical elements.
  - Individually.
  - In combination.

Operational Analysis

You also need to understand where you are in the survivability spectrum right now; do you have the resources you need?

- Assess your current capability to survive threats.
  - Personnel current capabilities.
  - Personnel current authorities.
  - Current physical capabilities.
  - On-call physical capabilities.

Survival Planning

Once you know where you have gaps, you must create ways to at least bridge them (even if you can't entirely fill them in).

- Develop specific alternatives that de-risk any critical elements.
- Immediate fixes.
- Network evolution.
- Define your implementation of continuity.
  - Structure (on-call versus routine topology).
  - Operations.

- Devolution of authority.
- Preestablished procedures.
- Practice.

Reality Check

Nice ideas are just that: nice. What matters is will your ideas actually work?

Evaluate your current survivability plan.

- Can you trust your people and their ingenuity?
    - On-call backup facilities.
    - Are they in fact sufficient to cover your identified gaps?
    - Will they be there when needed (the status of the contracted provider)?
    - Internal redundant structures.
    - Will they be implemented when needed?
    - Is reconfiguration at other sites required?
        - How will they know?
    - Who has authority to order reconfiguration?
        - Is there a succession chain on that authority?
    - Will customers/partners/vendors still be able to reach you?
    - Will phone numbers roll over?
    - Will email be rerouted?
    - Who will notify external parties?
    - On whose authority?

Recovery

Surviving past the immediate moment is only a beginning. You must be able to pick up and get going again (or, better yet, get back up to speed after having only slowed down instead of being stopped in your tracks).

- Do you want to return to your previous design or evolve toward a future design?

- Identify the location if relocation is required.
- Estimate your cyber-attack reconfiguration timelines.
  - Compromised system identification.
  - Disinfection.
  - Validation before returning to connectivity with the network.
- Estimate your physical loss relocation timelines.
  - Location acquisition.
  - Utilities, communications connected and configured.
  - Internal network installed and configured.
  - Intranet connectivity validated and redundancy configurations validated.
- Notify your business contacts (partners/vendors/customers).

APPENDIX D

Post-Mortem Questions

After every exercise and, more importantly, after a real disaster, you must gather information and sit down to consider what happened versus what you expected to happen, what worked versus what didn't, and so forth. You shouldn't do this the very next day, as there will be a great deal of information that takes some time to develop. On the other hand, you should not get too far away from events, or memories may fade and/or clash, and it then becomes difficult to find the nuggets of truth you need.

This is discussed in Chapter 10, "Returning from the Wilderness;" my personal rule of thumb is one week after things are stable—but stable is a qualitative judgment you must make, based on your knowledge of your network.

- How close was the event to what you envisioned when you made your plan?
  - Evaluate the significance of differences.
- How well did the plan work?
  - Assume perfect implementation.
  - Evaluate the plan's inherent quality.

- How well did you implement it?
    - Faithfulness of your execution.
    - Effectiveness of your execution.
    - Efficiency of your execution.
- What did you not expect that you should have?
    - Did any unexpected event violate the plan's underlying assumptions?
    - Why did you not anticipate this?
    - Its effect on the validity of the plan (re-evaluate).
- Where did you get lucky?
    - Based only on the actual occurrences.
    - Why was the impact less than it could/should have been for each?
- What could have happened (but didn't) that would have thrown everything askew?
    - Estimate your reaction, based on your knowledge available at the time.
    - What should you have done (with perfect hindsight)?
    - Its effect on the validity of the plan (re-evaluate).
- What did you as management do that helped? And what hurt?
    - Its effect on the validity of the plan (re-evaluate).

Time Value of Money

In presenting your business case, it is important to compare costs and benefits in dollars (or euros, or yen) of the same value. Because the value of money changes over time, you must correct for that. You must put all prices, costs, and revenues in dollars measured at the same time. When working with costs over a period of time, the costs are usually brought back to present value (PV), the costs valued in this year's dollars. When projecting an expense or income that will occur some time in the future, you may want to take the revenues obtained or costs incurred previously forward to compare them against the future value (FV) at that time. Either direction works because they use the same formula: $FV = PV(1 + i)^t$, where i is the interest rate used to inflate or deflate the values and t is the number of periods over which the value change occurs (usually measured in years).

To use the example from the discussion in Chapter 11, "The Business Case," we are planning to make an investment of $10,000 per year for five years. At the end of that period, the investment will start improving revenue. What improvement of revenue in the sixth year would cause the (nominal) $50,000 investment to break even? We will use an interest rate of 6.00 percent. Calculating the future value of each $10,000 investment, we have a total value by the sixth year of $59,753. That is the break-even point for this project.

Table E.1 Inflating to Find Future Value

Year	1	2	3	4	5	6
1	**$10,000**	$10,600	$11,236	$11,910	$12,625	$13,382
2		**$10,000**	$10,600	$11,236	$11,910	$12,625
3			**$10,000**	$10,600	$11,236	$11,910
4				**$10,000**	$10,600	$11,236
5					**$10,000**	$10,600
					Total →	$59,753

For instance, $10,000 spent this year inflated to its value in the sixth year is $FV = 10000(1.06)^5$—the exponent is 5 because there are five compounding periods between this year and the sixth year. This is shown progressing through the years in Table E.1.

We can also deflate each year's expense back to the current year in order to plan how much this will cost in today's dollars. For that, we will invert the formula to solve for PV: $PV = FV/(1 + i)^t$. The deflated value (or present value) of the five nominal $10,000 investments is $44,651.

For instance, from the third year back to the present, there are two compounding periods, so we use $PV = 10000/(1.06)^2$. That is the value this year of $10,000 in year 3, based on an interest rate of 6.00 percent. The walk-back from a given year to the present value is shown in Table E.2.

Table E.2 Deflating to Find Present Value

Year	1	2	3	4	5
1	**$10,000**				
2	$9,434	**$10,000**			
3	$8,900	$9,434	**$10,000**		
4	$8,396	$8,900	$9,434	**$10,000**	
5	$7,921	$8,396	$8,900	$9,434	**$10,000**
	$44,651	← Total			

Glossary

ARIN American Registry of Internet Numbers. Assigns numbers for the Americas, the Caribbean, and Sub-Saharan Africa.

ATM Asynchronous Transfer Mode. A cell-based WAN technology.

CLEC Competitive Local Exchange Carrier. A relatively new entrant to the local telephone service market (see ILEC).

CO Central Office. The telephony switching center that connects end lines into the PSTN.

CPE Customer Premises Equipment. Hardware located at the customer's site; may be leased to the customer by the provider or be provided by the customer (leading to alternative definition of Customer Provided Equipment). Configuration responsibility negotiable if service-firm provided.

CPX Command Post Exercise. An exercise limited to decision makers, intended to hone their skills and judgment. All those with whom they must interact are simulated. An inexpensive means to evaluate plans.

Demarc   Demarcation point. Where operational responsibility for service changes from one organization to another.

DMZ   Demilitarized Zone. An area accessible to the public, with servers providing information you wish the public to be able to see. No proprietary or internal and confidential information is ever supposed to be present here; it is not safe because it is not inside the firewall.

DNS   Domain Name System. The protocols by which human-friendly names can be used to locate resources on a TCP/IP network. Alternatively, Domain Name Server, the host providing name resolution to the numerical IP address.

DSCP   DiffServ (Differentiated Services) Code Point. A setting in the IP header to enable assignment of next-hop behavior.

EMP   Electromagnetic Pulse. A surge of energy that, if sufficiently powerful can damage or even destroy electronic equipment.

IANA   Internet Assigned Numbers Authority. The group that assigns number designators to protocols, ports, and so on. See URL in Appendix A, "References," under *Miscellaneous*.

iBGP   Internal BGP (Border Gateway Protocol). BGP is a traffic management protocol; iBGP is used inside one Autonomous System (a network with a single administration).

ILEC   Incumbent Local Exchange Carrier. The original monopoly telephone provider or its successor company (the Baby Bells, such as BellSouth and SBC, for instance).

IP   Internet Protocol.

IPv4   IP version 4. The current version in global use. Will eventually be superseded by version 6 (IPv6), probably sometime in the next decade.

IS   Information Services. The organization responsible for organizing, maintaining, and delivering information throughout the company.

ISP Internet Service Provider. A business that provides connectivity and other services to others desiring access to the global Internet.

LAN Local Area Network. A network within one area; originally a small network, typically confined to a single building. This is becoming less useful, and the boundary between the LAN and the WAN is becoming more a function of when service between networks must be leased rather than internally-provided. This often sets the demarc at the metro boundary.

NAT Network Address Translation. A service, typically provided on high-performance routers, that translates one source or destination IP address on a packet to another. It is processor-intensive but offers some anonymity and facilitates the use of private IP address space.

NOC Network Operations Center. The location from which a network is managed. It typically has monitoring stations that receive message traffic concerning network element health and may use an operational surveillance system to collect and assist in the interpretation of performance statistics.

NSP Network Services Provider. A business providing external connectivity to other businesses. Services may include, but are not limited to, ISP service.

OAM Operations, Administration, and Maintenance. Traffic within the WAN that is used to manage the WAN network. It is the highest-priority traffic because it enables the WAN to adapt to congestion, outages, and so on.

PBX Private Branch Exchange. A telephone switching system to manage a large number of lines in a business setting.

Port A physical and/or logical interface on a networking device (router or switch), through which data passes. Alternatively, a process or application identifier distinguishing the traffic stream of one application from another; found in the TCP and UDP headers. Alternatively, to migrate software from one OS to another. Usage is contextual.

Private IP Space   IP address blocks that should never be advertised to the public (beyond their own network). Because they are unknown to the outside world, they may be used repeatedly, by many networks, for internal addressing.

PSTN   Public Switched Telephone Network. The original, basic telephone service network. A great deal of WAN traffic is carried on its excess capacity; voice traffic, however, will always have the highest priority after OAM traffic.

SLA   Service Level Agreement. A contract between the service provider and the customer, specifying the terms of service (traffic carriage parameters) and obligations of each side. It typically includes penalties, but these are relatively modest.

TCP   Transmission Control Protocol. The protocol that governs reliable traffic transfer over IP.

UDP   User Datagram Protocol. A connectionless protocol operating over IP; information transfer controls are left to other protocols, making UDP data transfers faster and with lower overhead, but also inherently less reliable.

UPS   Uninterruptible Power Supply. A device to (at least) provide power long enough to allow a graceful system shutdown in the event of power loss from the utility. At best, you can continue to operate for quite some time on the UPS's battery. Performance varies widely according to type and capacity. *Caveat emptor.*

URL   Universal Resource Locator. The recognizable human-friendly name that DNS can translate into an IP address. The leading characters designate the protocol being used (example: http://www.wiley.com— http is the hypertext transfer protocol, used for Web pages).

VLAN   Virtual LAN. A form of logical grouping of hosts that does not require them to be on the same physical network (they need not share the same wire, as in an ordinary LAN). Used to segregate users and their traffic from one another.

WAN Wide Area Network. A network connecting other networks across gaps they cannot bridge with internal systems. Originally significantly slower than internal networks, this is rapidly evolving at the same time the demarc between the LAN and the WAN is becoming more fluid.

Index

NUMBERS

2001 CSI/FBI Computer Crime and Security Survey (Richard Power), 276

A

access lists, router configuration, 76
accounts, weak/no passwords, 62–63
addresses, egress/ingress, 64
Address Resolution Protocol (ARP), 43
Administrator account, Windows NT/2000, 41
admins, UNIX systems, 41
algorithms
  Code Excited Linear Predictive (CELP), 102
  encryption, 30
Allen, Julia H. (*CERT Guide to System and Network Security Practices*), 275
American Registry of Internet Numbers (ARIN), 293
anonymous logon, null session connection vulnerability, 67

anthrax attacks, 151–152
Application Layer, TCP/IP, 37–38
ARIN (American Registry of Internet Numbers), 293
ARPANET, worm development history, 48
Asynchronous Transfer Mode (ATM), 293
AT&T, service-affecting outages, 174–175
ATM (Asynchronous Transfer Mode), 293
attitude, natural disaster recovery, 97–98
authentication, SNMP vulnerability, 70
automated ping scripts, 44

B

Back Orifice Trojan, 49
backups
  colocated facilities, 118
  configuration file documentation, 127
  fraud/theft attack target, 30

299

backups *(continued)*
  network structure documentation, 127
  nonexistent/incomplete, 63
  secondary systems, 116–117
  Secured Storage, Inc., 129
backward compatibility, disadvantages, 266
bad luck (unfortunate opportunities)
  bird strikes, 171, 173
  route diversity, 175
  service-affecting outages, 172–175
Berkeley Internet Name Domain (BIND), UNIX system vulnerabilities, 68–69
Berkowitz, Howard (*WAN Survival Guide*), 183
biological attacks, 149–152
biowar hacker, anthrax attack, 152
bird strikes, bad luck (unfortunate opportunities), 171, 173
blitzkrieg analogy, cyber-attacks, 23–25
Blitzkrieg Story, The (Charles Messenger), 275
blood agents, 150
Boeing Aircraft, bird strike testing, 173
bombs, physical attack type, 147
books, resources, 275–276
BOOTP protocol, 41
boundaries
  bounded *vs.* unbounded networks, 15–18
  demarc (demarcation point), 14
bounded networks, *vs.* unbounded, 15–18
Bremer Commission, terrorism goals, 144
Britannic, sister ship to the *Titanic*, 88

British Board of Trade, lifeboat requirements, 88–89
budget
  direct costs, 245–246
  explicit costs, 247
  fixed costs, 244–245
  implicit costs, 247
  indirect costs, 245–246
  redundancy considerations, 120–121
  thought experiments, 6–7
  variable costs, 244–245
buffer overflows
  described, 62
  ISAPI extension, 66
  LPD (Line Printer Daemon), 69
  RPC services, 68
bugs, system element, 59–60
business case
  alternatives, 256
  alternatives summary presentation, 259–260
  CDG, Inc. data services presentation, 255–262
  cost-benefit analysis, 243
  direct costs, 245–246
  disaster summary presentation, 256–259
  expected values, 250–251
  explicit costs, 247
  fixed costs, 244–245
  implicit costs, 247
  indirect costs, 245–246
  present/future value calculations, 248
  presentation guidelines, 252–255
  proving ideas to skeptical audience, 243–244
  revenues, 249–250
  unmitigated risks presentation, 260–262

Index 301

valid comparisons, 247–249
variable costs, 244–245
business continuity
  CDG, Inc. data services plan, 130
  cyber-attacks, 5–6
  importance of, 2–3
  natural disasters, 4–5
business investments, trophy properties, 90–92

C

cache, ARP, 43
Californian, *Titanic* disaster warning, 89
Campbell, Stuart (*Security Transformation: Digital Defense Strategies to Protect Your Company's Reputation and Market Share*), 275
Canada, ice storms, 102–103
capability, operational continuity element, 117–118
Carpathia, *Titanic* mayday responder, 90
catastrophes
  predicting, 267–268
  statistics, 252
CBR (chemical/biological/radiological) attacks, physical attacks 149–152
CDG, Inc. data services
  alternatives summary presentation, 259–260
  backup increment, 129
  backup personnel, 138
  business case alternatives, 256
  business case presentation, 255–262
  business continuity plan, 130
  carrier activation, 129–130
  configuration control, 136–137
  connectivity requirement bundling, 129
  conspiracy of silence example, 182–186
  continental topology, 139–140
  development history, 129
  disaster summary presentation, 256–259
  facility lessons, 136
  lessons learned, 135–137
  logical topology, 130, 237–240
  master corporate plan, 130
  network evolution conclusions, 241–242
  potentially learned lessons, 138
  redundancy, 129
  scenario positives, 138–139
  Secured Storage, Inc. arrangement, 129
  topology lessons, 136
  total loss assumptions, 139–141
  UPS battery life considerations, 132–134
  UPS lessons, 137
  Virtual Private Networks (VPN), 129
  VoIP, 129
  winter storm affect, 131–132
CDs, cyber-recovery uses, 222
CELP (Code Excited Linear Predictive) algorithm, 102
Central Office (CO), 293
CERT Guide to System and Network Security Practices (Julia H. Allen), 275
chemical attacks, 149–152
Chicago, IL, man-made flood damage, 106–107
choking agents, 150
CLEC (Competitive Local Exchange Carrier), 293
coastal zones, major storm effects, 103–104

CO (Central Office), 293
Code Excited Linear Predictive (CELP) algorithm, 102
Code Red II worm, 48
Code Red worm, 48, 232
codes, DiffServ Code Point (DSCP), 42–43
collapsible lifeboats, *Titanic*, 89
Comdisco, Inc., disaster recovery business, 155
Command Post Exercise (CPX), 212, 293
Common Gateway Interface (CGI), system vulnerability, 65
Common Vulnerabilities and Exposures (CVE) project, 60
communications
  disaster planning, 215–217
  natural disasters, 93–94
comparisons, business case, 247–249
Competitive Local Exchange Carrier (CLEC), 293
complexity, system element, 58
configuration control, CDG, Inc. data services, 136–137
configuration files, documentation, 127
continental topology, CDG, Inc. data services, 139–140
continuity
  importance of, 2–3
  network survival requirements, 195–202
  operational, 116–126
  survivability choices, 213
  trade-offs, 270–274
continuity planning
  network requirements, 185
  operational analysis, 286
  reality *vs.* ideal plans, 287
  recovery plans, 287–288
  survival element, 192, 194
  survival planning, 286–287
  threat analysis, 286
converged networks, 82–85
corrupted services, quarantining, 270–271
costs
  direct, 245–246
  explicit, 247
  fixed, 244–245
  implicit, 247
  indirect, 245–246
  network protection, 78–80
  redundancy considerations, 120–121
  standardization benefits, 81
  trade-offs, 273–274
  variable, 244–245
CPE (Customer Premises Equipment), 293
CPX (Command Post Exercise), 293
CQD...MGY, *Titanic* distress call, 89
crackers
  described, 20
  vs. hackers, 19–20
credit cards, fraud/theft attacks, 29–30
Cuckoo's Egg, The: Tracking a Spy Through the Maze of Computer Espionage (Cliff Stoll), 276
Customer Premises Equipment (CPE), 293
customers, redundancy activation external notification, 123
cyber-attacks
  automated attack tools, 55–56
  automated ping scripts, 44
  blitzkrieg analogy, 23–25
  buffer overflows, 62
  business case presentation, 253–255

combining with physical attacks, 168–169
cyber-kidnapping, 166
cyber-recovery, 220–226
Denial of Service/Distributed DoS (Dos/DDoS), 49–51
Domain Administrator, 41
extortion, 33, 167
fraud/theft, 29–30
hacktivist, 21–22
immature (inexperienced) hackers, 13–19
industrial espionage, 27–29
infrastructure, 167–168
journeymen (deliberate) attackers, 19–26
logic bombs, 49
master attackers, 26–33
necessity of defense, 264
netspionage, 28
network damage potentials, 22
phone crackers (phrackers/phone phreaks), 84–85
ping floods, 50
point of origin statistics, 26
probability of, 5–6
probes, 43–45
reasons for, 11–12
record alterations, 31–33
recovery exercise, 234–235
root privilege, 41
sample network attack, 51–56
script kiddies, 21
sniffers, 45
spoofing, 30, 44
TCP floods, 50
terrorism, 166–168
threat analysis, 8–9, 204–206
time bombs, 49
Trojan horses, 48–49
UDP floods, 50
viruses, 45–46
war dialer, 44
war driving, 44
worms, 46–48
cyber-kidnapping, 166
cyber-recovery
  CDs, 222
  documentation, 222
  forensic procedures, 221–226
  information collection, 221–222
  logbooks, 222
  operational procedures, 220–221
  personnel assignments, 224
  response teams, 220–221
  risk tolerance, 225–226
cyber threats, analysis, 204–206
cyclones, adequate warning category, 99–104

D

daemons
  BIND, 68–69
  Line Printer Daemon (LPD), 69
  mountd, 69
  RPC, 68
  sadmind (System Admin Daemon), 69
damage potentials, 22
data currency, disaster planning element, 217
data integrity, importance of, 31–33
data traffic, *vs.* voice traffic, 83
default installations, general vulnerability type, 61–62
defenses
  basic, 265–266
  difficulties, 266–267
  reasons for, 264–265
demarc (demarcation point)
  defined, 294
  network boundary, 14

Demilitarized Zone (DMZ), 294
Denial of Service/Distributed DoS (Dos/DDoS) attacks, 49–51
Destination Address field, IP header, 42
DiffServ Code Point (DSCP) codes, 42–43, 294
DiffServ field, IP header, 42
direct revenues, 249–250
dirty bombs, radiological attack, 152
disaster planning
  command post exercise (CPX), 212
  communications, 215–217
  continuity, 192, 194
  cyber-threats, 204–206
  data currency, 217
  network continuity critical elements, 201–202
  network continuity service decomposition, 197
  network survival requirements, 195–202
  operational analysis, 206–207
  physical threats, 202–204, 205
  redundant facility locations, 214–215
  survivability in today's world, 213–217
  survival defined, 191–194
  survival planning, 207–212
  survival requirements, 194–207
  taxable *vs.* accounting profit, 193
  threat analysis, 202–206
  trade-offs, 218
  Web sites, 276
disaster summary, business case presentation element, 256–259
disasters, predicting, 257–268
discounting calculations, 248
DLLs (Dynamic Linked Libraries), 66

DMZ (Demilitarized Zone), 294
DNS (Domain Name System), 294
DNS servers, unfortunate planning example, 179
documentation
  configuration files, 127
  cyber-recovery element, 222
  network structure, 127
  operational continuity element, 120
Domain Administrator attacks, Windows NT/2000, 41
Domain Name Service (DNS)
  BIND (Berkeley Internet Name Domain), 68–69
  name resolution, 180
Domain Name System (DNS), 294
domain server, defined, 41
DSCP (DiffServ Code Point), 294
Dupuy, Trevor (*A Genius for War: The German Army and General Staff*), 275
Dynamic Linked Libraries (DLLs), 66

E

earthquake hazards, Web sites, 276–277
earthquakes
  Kobe, Japan, 107–108
  Mexico City, 91–92
  New Madrid, Missouri, 91
  Richter Scale, 108
  U.S. Geological Survey statistics, 108–112
effective, *vs.* efficient, 203
egress/ingress address filtering, system vulnerability, 64
Electromagnetic Pulse (EMP)
  bombs, physical attack type, 147–148
  defined, 294

Ellison, R.J. *"Survivable Network Systems: An Emerging Discipline,"* 275
El Nino, weather phenomenon, 103
email headers, spammer alterations, 32
emergency services, operational continuity element, 123–124
emergent properties, system element, 59
encryption algorithms, shortcomings, 30
encryption, LM hash, 67
Enterprise System Connection (ESCON), 195
equipment
  present value calculations, 248
  unfortunate implementation, 186–187
evacuations, unintentional disasters, 178
evolution, CDG, Inc.
  data services conclusions, 241–242
  services topology, 237–240
exercise planning
  cyber-attacks, 234–235
  disaster recovery rehearsal, 231–236
  importance, 271
  physical attacks, 235–236
expected values, business case element, 250–251
expenses
  direct *vs.* indirect costs, 245–246
  explicit costs, 247
  implicit costs, 247
  overhead *vs.* fixed costs, 244–245
externalities, described, 33–34
external notification, operational continuity element, 123
external routers, 76
extortion
  cyber-attacks, 167
  master attacker goal, 33
extranets, unbounded network link, 15–16

F

facilities
  CDG, Inc. data services, 136
  colocated, 118
  operational continuity element, 117–118
  redundant locations, 214–215
ferroresonant standby UPS, 133
Fibre Connection (FICON), 195
fields
  Destination Address, 42
  DiffServ, 42
  IP header, 42
  Service Type, 42
  Source Address, 42
files, configuration, 127
file transfers, fraud/theft attack target, 30
filters
  egress/ingress address, 64
  ports, 76–77
  router access lists, 76
  Unicode vulnerability protection, 66
fires, unintentional disasters, 175–176
fixed costs, business case element, 244–245
fixes
  long-term, 208, 209–210
  quick, 208
  survival planning element, 207–210
flags, TCP header, 39

flat topology, 72–73
flexibility, operational continuity element, 118–119
floods
   Dos/DDoS attacks, 50–51
   major storm effect, 103–104
focus, terrorism, 144
folders, shared vulnerability, 66–67
forensic procedures
   CDs, 222
   cyber-recovery, 221–226
   documentation, 222
   information collection, 221–222
   logbooks, 222
   personnel assignments, 224
   questions, 223
   risk tolerance, 225–226
formulas, figuring present value (PV) and future value (FV), 291–292
Fort Worth, TX, tornados, 107
fraud/theft attacks, master attackers, 29–30
future/present value calculations, 248
future value (FV), money, 291–292

G

Genius for War: The German Army and General Staff, A (Trevor N. Dupuy), 275
goals
   restoration, 228–231
   terrorism, 143

H

hackers
   biowar, 152
   described, 20
   hacktivist, 21–22
   immature (inexperienced), 13–19
   journeymen (deliberate) attackers, 19–26
   master attackers, 26–33
   network damage potential, 22
   phone crackers (phrackers/phone phreaks), 84–85
   script kiddies, 21
   spoofing, 30
   testers, 18–19
   voyeurs, 14–18
   vs. crackers, 19–20
hacktivist, described, 21–22
hardware
   configuration file documentation, 127
   natural disaster survivability, 126–129
   network structure documentation, 127
   present value calculations, 248
   unfortunate implementation, 186–187
hazardous materials
   required evacuations, 178
   unfortunate planning example, 181–182
headers
   IP, 40, 42
   TCP, 38–39
   UDP, 39–40
Host-to-Host Layer, TCP/IP, 37
hub topology, 72–73
human behaviors, *Titanic* disaster, 90
hurricanes
   adequate warning category, 99–104
   U.S. National Weather Service tracking, 100–101

I

IANA (Internet Assigned Numbers Authority), 294
iBGP (Internal BGP), 294
icebergs, *Titanic* warning, 89, 92
IIS Lockdown tool, Unicode vulnerability protection, 66
IIS RDS exploit, Windows PC vulnerability, 66
ILEC (Incumbent Local Exchange Carrier), 294
immature (inexperienced) hackers
  commonality of purpose, 13–14
  testers, 18–19
  voyeurs, 14–18
immediate operations, physical recovery, 226–227
Incumbent Local Exchange Carrier (ILEC), 294
indirect revenues, 249–250
industrial espionage, master attackers, 27–29
inflation calculations, 248
information collection, cyber-recovery element, 221–222
Information Services (IS), 294
infrastructure, cyber-attack target, 167–168
ingress/egress address filtering, system vulnerability, 64
ingress routers, 75–76
installations, general vulnerability type, 61–62
interaction, system element, 58–59
interest calculations, 248
Interexchange Carriers (IXCs), service-affecting outages, 174
Internal BGP (iBGP), 294
internal notification, operational continuity element, 123
Internet Assigned Numbers Authority (IANA), port designations, 39, 40, 294
Internet Control Message Protocol (ICMP), 42
Internet Protocol (IP), 294
Internet Relay Chat (IRC), flood attacks, 51
Internet Service Provider (ISP), 295
Internet Services Application Programming Interface (ISAPI), 66
Internet, unbounded network link, 15–16
Internetwork Layer, TCP/IP, 37
IP (Internet Protocol), 294
IP address
  backup servers, 123
  Network Address Translation (NAT), 199
IP headers, 40, 42
IP version 4 (IPv4), 294
IS (Information Services), 294
ISP (Internet Service Provider)
  defined, 295
  STL Network Services, Inc., 198

J

jitter, defined, 83
journeymen (deliberate) attackers
  attention seeking, 19
  blitzkrieg analogy, 23–25
  hacktivist, 21–22
  network damage potential, 22
  point of origin statistics, 26
  purpose of, 22–23
  script kiddies, 21

K

keystrokes, reasons for capturing, 47
Kobe, Japan, earthquake, 107–108

L

lahars, Mt. Rainier, 112
LAN (Local Area Network), 295
LAN Manager, LM hash vulnerability, 67
landslides, Mt. Saint Helens, 111–112
latency, defined, 83
Layer 3 (Network Layer), TCP/IP, 37
Layer 4 (Transport Layer), TCP/IP, 37
layers, TCP/IP, 36–37
LDAP, port 389, 77
legal actions, recovery considerations, 229
Lehman Brothers, WTC attack successes, 158–159
lifeboats, *Titanic*, 89
life cycles, redundant operation mode, 121–122
line-interactive UPS, 133
Line Printer Daemon (LPD), UNIX system vulnerabilities, 69
links, unbounded network, 15–16
LM hash, Windows PC vulnerability, 67
Local Area Network (LAN), 295
local loops, WTC attack, 162–164
logbooks, cyber-recovery element, 222
logging, nonexistent/incomplete, 64–65
logical topologies
  CDG, Inc. data services, 130, 237–240
  described, 73–75
logic bombs, 49
logons, anonymous, 67
long-term fixes, survival planning considerations, 208, 209–210
lost access, WTC attack, 159–161

M

Magic Lantern, keystroke-capturing software, 47
master attackers
  described, 26–27
  extortion, 33
  fraud/theft, 29–30
  industrial espionage, 27–29
  netspionage, 28
  record alteration, 31–33
  social engineering, 26
master corporate plan, CDG, Inc. data services, 130
McCarthy, Mary Pat (*Security Transformation: Digital Defense Strategies to Protect Your Company's Reputation and Market Share*), 275
mesh topology, 72–73
Messenger, Charles (*The Blitzkrieg Story*), 275
Metromedia Fiber Network Services, Inc. (MFN), WTC attack losses, 161
Mexico City, Mexico, earthquake, 91–92
Microsoft, DNS server unfortunate planning example, 179
money, figuring value, 291–292
Morris, Robert, worm development history, 48
mountd, UNIX systems vulnerabilities, 69
mstream, automated attack tool, 55
Mt. Pinatubo, volcanic eruption, 110–111
Mt. Saint Helens, volcanic eruption, 111
Munich: The Price of Peace (Telford Taylor), 276

Index   309

N

name resolution, 180
name-resolution accessibility, operational continuity element, 117–118
name servers, flood attacks, 50–51
NAT (Network Address Translation), 198, 199, 295
National Infrastructure Protection Center (NIPC), Top Twenty list, 60
natural disasters
  adequate warning plans, 99–104
  attitude, 97–98
  CDG, Inc. data services example, 129–141
  Chicago River flood, 106–107
  communications, 93–94
  earthquakes, 107–108
  El Nino phenomenon, 103
  lahars, 112
  landslides, 111–112
  lessons from successes, 94–99
  major storm effects, 103–104
  Mexico City earthquake, 92
  modest warning category, 105–107
  network assets, 126–129
  New Madrid, Missouri earthquake, 91
  no-notice event types, 128
  operational continuity, 116–126
  organization, 95–96
  planning, 98–99
  predicting, 367
  probability of, 4–5
  real estate considerations, 127–128
  Richter Scale, 108
  scarcity of heroes, 94
  storm warning quality issues, 100
  sudden storms, 107–112
  Titanic, 88–94
  training for, 92–93, 96–97

Tunnel Fire, 105–106
volcanoes, 110–112
warning acknowledgement, 92
winds, 104
negative externalities, described, 34
nerve agents, 151
NetBIOS Session Service, null session vulnerability, 67
NetBIOS, unprotected Windows networking shares, 66–67
netspionage, defined, 28
Network Address Translation (NAT), 198, 199, 295
network blackjack, Port 1025 assignment, 76–77
Network Interface Layer, TCP/IP, 37
Network Layer, TCP/IP, 37
Network Operations Center (NOC), 295
network requirements, continuity planning, 185
networks
  bounded *vs.* unbounded, 15–18
  converged, 82–85
  documentation structure, 127
  protection costs, 78–80
  standardization benefits, 80
  Telephony, 84–85
  topologies, 72–75
Network Services Provider (NSP), 295
New Madrid, Missouri, earthquake, 91
New York Board of Trade (NYBOT), WTC attack successes, 154–155
New York City OEM, WTC attack, 164–165
Nimda worm, 48
NIPC (National Infrastructure Protection Center), 60

NOC (Network Operations Center), 295
NSP (Network Services Provider), 295
null sessions
  information leakage vulnerability, 67
  NetBIOS Session Service vulnerability, 67
NYBOT (New York Board of Trade), WTC attack successes, 154–156

O

Oakland, CA, Tunnel Fire disaster, 105–106
OAM (Operations, Administration, and Maintenance), 295
off-site personnel, operational continuity importance, 124–125
Oklahoma City, OK, tornados, 107
Olympic, sister ship to the *Titanic*, 88
one-for-all redundancy, 118
one-for-many redundancy, 118
one-for-one redundancy, 118
online power UPS, 133–134
on-site personnel, operational continuity importance, 125–126
operational analysis
  continuity, 286
  disaster planning, 206–207
operational continuity
  colocated facilities, 118
  emergency services, 123–124
  external notification, 123
  flexibility, 118–119
  internal notification, 123
  internal notification, 123
  key people identification, 119–120
  name-resolution accessibility, 117–118
  off-site personnel, 124–125
  one-for-all redundancy, 118
  one-for-many redundancy, 118
  one-for-one redundancy, 118
  on-site personnel, 125–126
  questions, 117–123
  redundancy benefits, 116–120
  redundant mode life cycle, 121–122
  standardized documentation, 120
operational procedures, cyber-recovery, 220–221
Operations, Administration, and Maintenance (OAM), 295
operator error, system vulnerability, 85–86
organization, natural disaster recovery, 95–96
OSI Reference Model, TCP/IP layers, 36–37
outages, service-affecting, 172–175
overhead expenses, *vs.* fixed costs, 244–245
overseas operations, benefits, 263–264

P

partners, redundancy activation external notification, 123
passwords
  alternate location access considerations, 127
  LM hash vulnerability, 67
  system account vulnerability, 62–63
patches
  BIND daemon, 69
  LPD (Line Printer Daemon), 69
  mountd, 69
  sadmind (System Admin Daemon), 69
  sendmail vulnerabilities, 68
  UNIX system RPC daemons, 68

Windows 2000 Service Pack 2, 66
Windows NT 4.0 Security Roll-Up
    Package, 66
PBX (Private Branch Exchange), 295
performance, wrong problem
    solving example, 188–190
personnel
  CDG, Inc. data services
      considerations, 138
  forensic evidence responsibilities,
      224
  key people identification, 119–120
  logbooks, 222
  off-site, 124–125
  on-site, 125–126
  redundancy activation internal
      notification, 123
Petronas Towers, trophy property,
    91
phone crackers, Telephony
    networks, 84–85
physical attacks
  anthrax, 151–152
  biological, 149–152
  blood agents, 150
  bombs, 147
  business case presentation, 253–255
  CBR (chemical/biological/
      radiological) attacks, 149–152
  chemical, 149–152
  choking agents, 150
  combining with cyber-attacks,
      168–169
  dirty bombs, 152
  electromagnetic pulse (EMP),
      147–148
  nerve agents, 151
  physical recovery, 226–228
  radiological, 149–152
  recovery exercise, 235–236

sabotage, 148–149
skin agents, 150
tear gas, 150
Physical Layer, TCP/IP, 37
physical recovery
  immediate operations, 226–227
  sustained operations, 227–228
physical threats, analysis, 202–204,
    205
physical topologies, 72–73
ping (packet internet groper), 42
ping floods, Dos/DDoS attack, 50
planning, unfortunate, 178–186
plans
  adequate warning, 99–104
  natural disaster recovery, 98–99
point of origin, cyber-attack
    statistics, 26
port 53 (TCP/UDP), 77
port 389 (LDAP), 77
port 1025, network blackjack, 76–77
ports
  commonly abused, 78
  defined, 295
  filtering, 76–77
  IANA designations, 39, 40
  LDAP (389), 77
  NetBIOS null session, 67
  network blackjack (1025), 76–77
  showing assigned, 39
  system vulnerability, 63–64
  TCP header, 38–39
  TCP/IP, 38
  TCP/UDP (53), 77
  well-known, 41
port scanners, 64
positive externalities, described, 34
post-mortem questions, 249–250
Powers, Richard (*2001 CSI/FBI
    Computer Crime and Security
    Survey*), 276

Powers, Richard (*Tangled Web: Tales of Digital Crime from the Shadows of Cyberspace*), 167, 275
presentations
　alternatives summary, 259–260
　business case, 252–255
　disaster summary, 256–259
　unmitigated risks, 260–262
present/future value calculations, 248
present value (PV), money, 291–292
Private Branch Exchange (PBX), 295
private IP space, defined, 296
probes
　automated ping scripts, 44
　described, 43–45
　sniffers, 45
　spoofing, 44
　war dialer, 44
　war driving, 44
procedures, survival planning, 211–212
Process of Network Security: Designing and Managing a Safe Network (Thomas Wadlow), 276
productivity, lost contributions, 232
profit, taxable *vs.* accounting, 193
protocols
　ARP, 43
　BOOTP, 41
　ICMP, 42
　porous, 266
　Server Message Block (SMB), 66–67
　Simple Network Management Protocol (SNMP), 70
　TCP (Transport Control Protocol), 38–40
　TCP/IP, 36–43
　UDP (User Datagram Protocol), 38–40
Public Switched Telephone Network (PSTN), 296

Q

quarantine team, cyber-recovery, 221
questions
　cyber-recovery forensic procedures, 223
　disaster planning, 281–283
　operational continuity, 117–123
　post-mortem examinations, 229–231, 289–290
quick fixes, survival planning concerns, 208
Qwest, service-affecting outages, 172, 175

R

r commands, UNIX system vulnerabilities, 69
radiological attacks, 149–152
real estate, natural disaster considerations, 127–128
reality *vs.* ideal plans, continuity, 287
record alterations, master attacker goal, 31–33
recovery
　cyber-attack rehearsal, 234–235
　exercise planning, 231–236
　legal action considerations, 229
　lost revenue contributions, 232
　network evolution planning inclusion, 236–242
　physical attack rehearsal, 235–236
　physical recovery, 226–228
　planning, 268–270
　post-mortem examinations, 229–231
　rehearsing, 231–236
　restoration process, 228–231
recovery planning, cyber-recovery, 220–226
recovery plans, continuity, 287–288

Red Hat Linux 5.2, security concept testing, 56
redundancy
 CDG, Inc. data services, 129
 cost considerations, 120–121
 facility locations, 214–215
 key people identification, 119–120
 life cycles, 121–122
 one-for-all, 118
 one-for-many, 118
 one-for-one, 118
 operational continuity element, 116–120
remedies, survival planning element, 210–211
Remote Data Service (RDS), Windows PC vulnerability, 66
Remote Procedure Calls (RPCs), UNIX system buffer overflows, 68
Republic of the Philippines, Mt. Pinatubo eruption, 110–111
resiliency, trade-offs with standardization, 272
resources, Web sites, 276–280
response teams, cyber-recovery, 220–221
revenues, business case element, 249–250
Richter Scale, earthquakes, 108
right-of-way permits, route diversity issues, 176–177
risk tolerance, cyber-recovery, 225–226
rodents, unintentional disasters, 183, 185
root privilege attacks, UNIX systems, 41
route diversity, right-of-way permits, 176–177

route diversity, unintentional disaster awareness, 175–177
routers
 access lists, 76
 defense in depth, 75–76
 external, 76
 ingress, 75–76
 logical topology element, 74–75

S

sabotage, physical attack type, 148–149
sadmind (System Admin Daemon), UNIX system vulnerabilities, 69
San Diego Supercomputer Center, Red Hat Linux 5.2 concept test, 56
San Francisco, CA
 earthquake probability statistics, 108–110
 storm damage, 105
SANS Institute, Twenty Most Critical Internet Security Vulnerabilities, 60
scanners, port, 64
script kiddies, described, 21
scripts
 automated attack tools, 55–56
 automated pings, 44
 script kiddies, 21
 war dialer, 44
Sears Tower, trophy property, 91
secondary systems
 network structure documentation, 127
 password access considerations, 127
 redundancy benefits, 116–120
Secured Storage, Inc., CDG, Inc. data services arrangement, 129

Security Transformation: Digital Defense Strategies to Protect Your Company's Reputation and Market Share (Mary Pat McCarthy and Stuart Campbell), 275
segments, TCP/IP, 38
sendmail, UNIX system vulnerabilities, 68
Server Message Block (SMB) protocol, 66–67
servers, backup IP address, 123
service-affecting outages, 172–175
Service Level Agreement (SLA), 296
Service Type field, IP header, 42
sessions, TCP/IP, 37–38, 41
shared folders, NETBIOS unprotected Windows networking shares, 66–67
Shipbuilder magazine, 88
signal rockets, *Titanic*, 89
Simple Network Management Protocol (SNMP), UNIX system vulnerability, 70
skin agents, 150
SLA (Service Level Agreement), 296
SMB (Server Message Block) protocol, 66–67
smurf attacks, automated attack tool, 55
sniffers, 45
social engineering, defined, 26–27
software developers, India, 264
Source Address field, IP header, 42
spammers, email header alteration, 32
spoke topology, 72–73
spoofing, 30, 44
Sprint, service-affecting outages, 172, 175
stacheldraht, automated attack tool, 55

standardization
  network benefits, 80–81
  trade-offs with resiliency, 272
standardized documentation, operational continuity element, 120
standby power UPS, 133
star topology, 72–73
STL Network Services, Inc.
  critical elements, 201–202
  cyber-threat analysis, 204–206
  ISP services, 198
  long-term fixes, 209–210
  mission definition, 196
  network continuity requirements, 195–202
  OAM traffic, 196–197
  operational analysis, 206–207
  physical threat analysis, 202–204
  port-based management VLAN, 198
  quick fixes, 208–209
  service decomposition, 197
  services, 195, 198
  survival planning, 207–212
  threat analysis, 202–206
  VPN services, 198, 200
  WAN services, 198, 200
Stoll, Cliff (*The Cuckoo's Egg: Tracking a Spy Through the Maze of Computer Espionage*), 276
SunGard, disaster recovery business, 155
superusers, UNIX systems, 41
"*Survivable Network Systems: An Emerging Discipline*" (R.J. Ellison), 275
survival
  continuity planning, 192, 194
  cyber-threats, 204–206
  defined, 3–4, 191–194

efficient, *vs.* effective, 203
network continuity critical elements, 201–202
network continuity requirements, 195–202
network continuity service decomposition, 197
operational analysis, 206–207
physical threats, 202–204, 205
requirements, 194–207
taxable *vs.* accounting profit, 193
threat analysis, 202–206
survival planning
 command post exercise (CPX), 212
 continuity, 286–287
 fixes, 207–210
 procedures, 211–212
 remedies, 210–211
sustained operations, physical recovery, 227–228
systemic behavior
 trade-offs, 270–272
 replacement elements behaviors, 271
systems
 accounts with weak/no passwords, 62–63
 anonymous logon vulnerability, 67
 BIND weaknesses, 68–69
 buffer overflows, 62, 68
 bugs, 59–60
 CGI program vulnerability, 65
 complexity, 58
 CVE project, 60
 default installation vulnerability, 61–62
 educational, 57–58
 emergent properties, 59
 general vulnerabilities, 61–65
 IIS RDS exploit, 66
 ingress/egress address filtering, 64

interaction, 58–59
 ISAPI extension buffer overflows, 66
 NETBIOS unprotected Windows networking shares, 66–67
 network topologies, 72–75
 nonexistent/incomplete backups, 63
 nonexistent/incomplete logging, 64–65
 open ports, 63–64
 operator error, 85–86
 port scanners, 64
 r commands, 69
 redundancy benefits, 116–120
 secondary, 116–117
 sendmail vulnerabilities, 68
 TANSTAAFL trade-offs, 58
 Twenty Most Critical Internet Security Vulnerabilities, 60–70
 Unicode vulnerability, 66
 UNIX vulnerabilities, 68–70
 vulnerability common threads, 70–71
 Windows PC vulnerabilities, 65–67

T

2001 CSI/FBI Computer Crime and Security Survey (Richard Power), 276
tables, ARP cache, 43
Tangled Web: Tales of Digital Crime from the Shadows of Cyberspace (Richard Powers), 167, 275
Taylor, Telford (*Munich: The Price of Peace*), 276
TCP (Transmission Control Protocol), 296
TCP floods, Dos/DDoS attack, 50
TCP headers, 38–39

TCP/IP
  Application Layer, 37–38
  ARP protocol, 43
  development history, 36
  DiffServ Code Point (DSCP) codes, 42–43
  flood attacks, 50–51
  goals, 36
  Host-to-Host Layer, 37
  IANA port designations, 39, 40
  ICMP, 42
  Internetwork Layer, 37
  IP headers, 40, 42
  layers, 36–37
  Network Interface Layer, 37
  Network Layer (Layer 3), 37
  Physical Layer, 37
  ports, 38
  segments, 38
  sessions, 37–38, 41
  TCP headers, 38–39
  Transport Layer (Layer 4), 37–38
  UDP header, 40
  well-known ports, 41
TCP/UDP, port 53, 77
tear gas, 150
Telephony networks, phone crackers, 84–85
terrorism
  biological attacks, 149–152
  bombs, 147
  CBR (chemical/biological/radiological) attacks, 149–152
  chemical attacks, 149–152
  combatant/noncombatant indifference, 144–145
  combined physical/cyber attacks, 168–169
  cyber attacks, 166–168
  cyber-kidnapping, 166
  defined, 144
  development history, 143
  electromagnetic pulse (EMP), 147–148
  emotional impact message, 144
  events spectrum, 145
  extortion, 167
  focus, 144
  goals, 143
  member population considerations, 145
  physical attack types, 146–152
  radiological attacks, 149–152
  sabotage, 148–149
  success rates, 143
  target values, 145
  Web sites, 278–279
  WTC (World Trade Center), 153–166
testers, immature (inexperienced) hackers, 18–19
theft/fraud attacks, master attackers, 29–30
thought experiments, budget protection, 7
threat analysis
  continuity, 286
  cyber-threats, 204–206
  physical threats, 202–204
threat assessments, pre-planning for disasters, 8–9
time bombs, 49
timelines
  post-mortem examinations, 229–231
  World Trade Center (WTC), 153
Titanic
  Californian iceberg warning, 89
  collapsible lifeboats, 89
  CQD...MGY distress calls, 89
  human behaviors, 90
  iceberg warnings, 89, 92
  lifeboats, 88–89

Index

scarcity of heroes, 94
sea trials, 91
signal rockets, 89
sister ships, 88
trophy property, 90–92
virtually unsinkable, 88
tools, automated attack, 55–56
topologies
   CDG, Inc. data services, 130, 136, 237–240
   continental, 139–140
   logical, 73–75
   physical, 72–73
tornados
   Fort Worth, TX, 107
   Oklahoma City, OK, 107
t0rnkit, automated attack tool, 56
traceroute, 42
trade-offs
   costs, 273–274
   standardization and resiliency, 272
   systemic behavior, 270–272
traffic, data *vs.* voice, 83
training
   natural disaster recovery, 96–97
   natural disasters, 92–93
Transmission Control Protocol (TCP), 296
Transmission Control Protocol/Internet Protocol (TCP/IP), 36–43
Transport Control Protocol (TCP), 38–40
Transport Layer, TCP/IP, 37–38
Tribal Flood Network 2000 (TFN2K), automated attack tool, 55
Tribal Flood Network (TFN), automated attack tool, 55
trinoo, automated attack tool, 55
Trojan horses
   Back Orifice, 49
   described, 48–49

trophy property, business investments, 90–92
tropical storm information, Web sites, 279
Twenty Most Critical Internet Security Vulnerabilities, 60–70
typhoons, adequate warning category, 99–104

U

UDP (User Datagram Protocol), 296
UDP floods, Dos/DDoS attack, 50
UDP headers, 39–40
unbounded networks
   fraud/theft attacks, 29–30
   vs. bounded, 15–18
undress rehearsals, disaster recovery, 231–236
unfortunate implementation, wrong problem solving, 188–190
unfortunate planning
   hazardous materials example, 181–182
   rodents, 183, 185
   theirs, 181–184, 186
   yours, 178–181
Unicode Standard, Windows system vulnerability, 66
unintentional disasters
   fires, 175–176
   required evacuations, 178
   rodents, 183, 185
   route diversity, 175, 176–177
   service-affecting outages, 172–175
   unfortunate implementation, 186–190
   unfortunate opportunities, 171–178
   unfortunate planning, 178–186
   wrong problem solving, 188–190
Uninterruptible Power Supply (UPS), 133–134, 296

Universal Resource Locator (URL), 66, 296
UNIX systems
 admins, 41
 BIND weaknesses, 68–69
 buffer overflows in RPC services, 68
 default SNMP strings, 70
 Line Printer Daemon (LPD), 69
 mountd, 69
 r commands, 69
 root privilege attacks, 41
 sadmind (System Admin Daemon), 69
 sendmail vulnerabilities, 68
 super users, 41
unmitigated risks, business case presentation, 260–262
unnatural disasters, predicting, 267–268
UPS (Uninterruptible Power Supply), 296
UPS capabilities, Web sites, 279
URL (Universal Resource Locator), 296
User Datagram Protocol (UDP), 38–40, 296
U.S. Geological Survey, earthquake statistics, 108–112
U.S. National Infrastructure Protection Center, 180
U.S. National Institute of Standards and Technology (NIST), CVE metabase, 60
U.S. National Weather Service, storm tracking, 100–101
U.S. Secret Service, WTC attack, 165–166

V

vaccination team, cyber-recovery, 221
variable costs, business case element, 244–245
vendors, redundancy activation external notification, 123
Verizon
 service-affecting outages, 173, 175
 WTC attack access losses, 160–161
Virtual LAN (VLAN), 296
Virtual Private Networks (VPN), CDG, Inc. data services, 129
viruses
 described, 45–46
 vs. worms, 47
VLAN (Virtual LAN), 296
voice traffic, *vs.* data traffic, 83
VoIP, CDG, Inc. data services, 129
volcanic eruption data, Web sites, 279
volcanoes, Mt. Pinatubo, 110
voyeurs
 bounded *vs.* unbounded networks, 15–18
 demarc (demarcation point), 14
 immature (inexperienced) hackers, 14–18
VPN services, STL Network Services, Inc., 198, 200
vulnerabilities, general, 61–65

W

Wadlow, Thomas (*Process of Network Security: Designing and Managing a Safe Network*), 276
Wall Street Journal, WTC attack successes, 156–157
WAN (Wide Area Network), 296

WAN services, STL Network Services, Inc., 198, 200
WAN Survival Guide (Howard Berkowitz), 183
war dialer, 44
war driving, 44
warnings
  adequate category, 99–104
  modest category, 105–107
  natural disasters, 99–112
  quality issues, 100
  sudden storm category, 107–112
weak passwords, system vulnerability, 62–63
weather
  adequate warning category, 99–104
  El Nino phenomenon, 103
  major storm effects, 103–104
  tornados, 107
  Web sites, 279–280
Web sites
  American Registry for Internet Numbers, 278
  Association of Bay Area Governments, 277
  automated scanner, 60
  Bremer Commission report, 278
  Centre for the Study of Terrorism and Political Violence, 278
  CERT, 277
  chemical/biological terror threat to US, 279
  damage estimates source, 278
  disaster planning firm, 277
  Disasters by Design, 276
  DNS server location problems, 277
  earthquake hazards, 276–277
  Earthquake hazards in Northern California/Nevada, 276
  EMP information, 279
  FCC outage reports, 277
  Global patterns of Terrorism, 278
  Highlights, 277
  IANA, 64
  International disaster situation reports, 276
  International Policy Institute for Counter-Terrorism, 278
  Internet Assigned Numbers Authority, 278
  Internet connectivity quality, 277
  Mitre, 60
  natural hazards cost, 278
  SANS Institute, 60
  terrorism, 278–279
  Terrorism Research Center, 278
  tropical storm information, 279
  UPS capabilities, 279
  U.S. Federal Emergency Management Agency, 276
  U.S. Geological Survey, 277
  U.S. National Geophysical Data Center, 276
  U.S. National Oceanic and Atmospheric Administration, 280
  U.S. National Weather Services, 280
  U.S. State Department report on global terrorism, 2000, 279
  volcanic eruption data, 279
  weather planning, 279–280
  World Trade Center and surroundings, real estate damage report 279
well-known ports, defined, 41
White Star Lines, trophy property, 90–92
Wide Area Network (WAN), 296
Windows 2000, Administrator account, 41
Windows 2000 Service Pack 2, Unicode vulnerability patch, 66

Windows NT
  Administrator account, 41
  Domain Administrator, 41
Windows NT/2000
  anonymous logon vulnerability, 67
  LAN Manager compatibility, 67
Windows NT 4.0 Security Roll-Up
    Package, ISAPI extension buffer
    overflow, 66
Windows PC
  anonymous logon vulnerability, 67
  IIS RDS exploit, 66
  information leakage via null
    session connections, 67
  ISAPI extension buffer
    overflows, 66
  LM hash, 67
  NetBIOS unprotected Windows
    networking shares, 66–67
  port assignments, 39
  system vulnerabilities, 65–67
  Unicode Standard vulnerability, 66
winds, major storm effect, 104
winter storms
  adequate warning category, 99–104
  Canadian ice storm, 102–103

World Trade Center (WTC)
  Lehman Brothers, 158–159
  local loops, 162–164
  lost access, 159–161
  Metromedia Fiber Network
    Services, Inc., 161
  New York City OEM, 164–165
  NYBOT (New York Board of
    Trade), 154–156
  parking garage vulnerability, 153
  successes, 154–159
  terror attack possibility reviews,
    153
  timelines, 153
  trophy property, 91
  U.S. Secret Service, 165–166
  Verizon, 160–161
  Wall Street Journal, 156–157
worms
  Code Red, 48
  Code Red II, 48
  described, 46–48
  Nimda, 48
  vs. viruses, 47